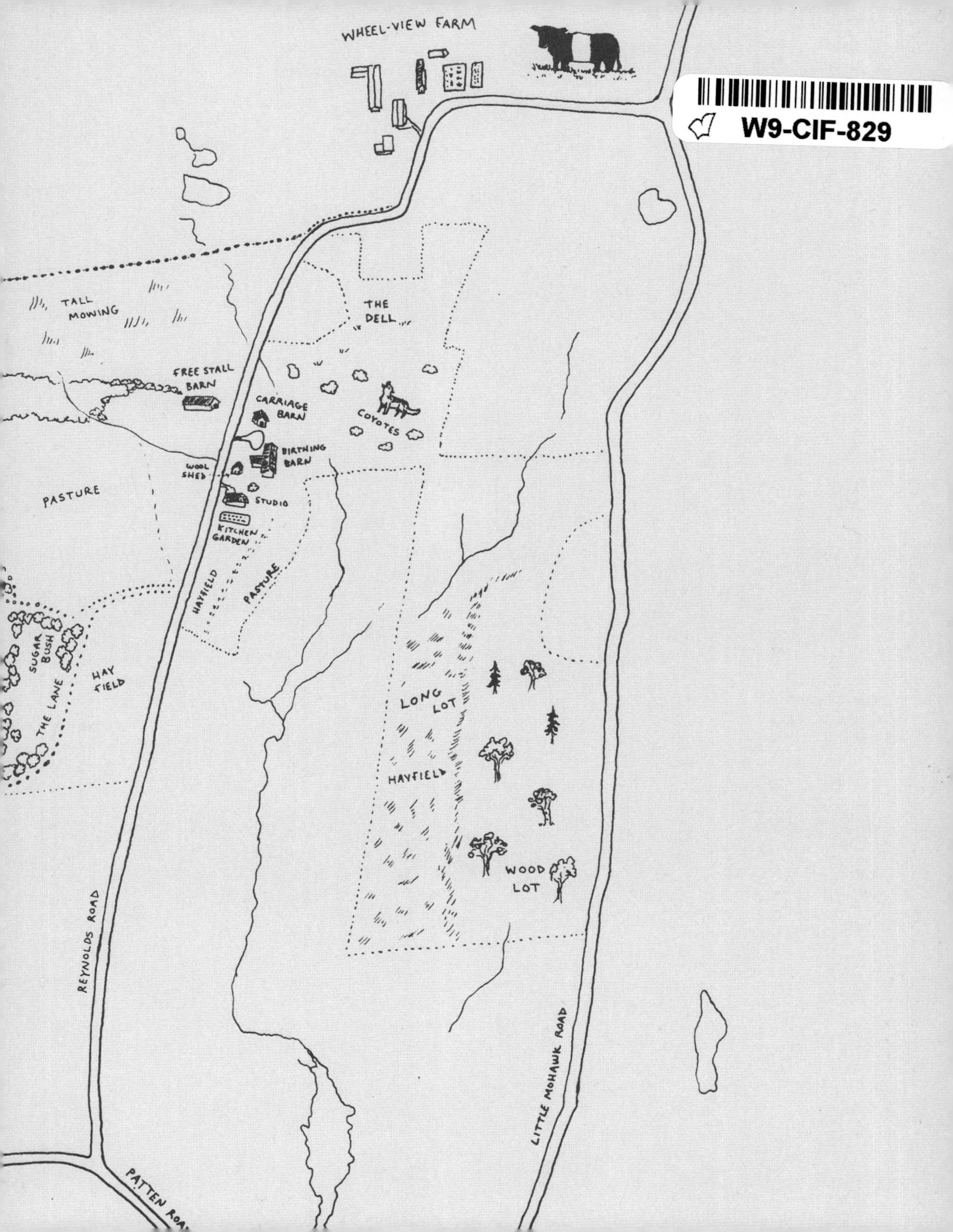
WHEEL-VIEW FARM
W9-CIF-829
TALL MOWING
THE DELL
FREE STALL BARN
CARRIAGE BARN
COYOTES
BIRTHING BARN
WOOL SHED
PASTURE
STUDIO
KITCHEN GARDEN
HAYFIELD
PASTURE
SUGAR BUSH
HAY FIELD
THE LANE
LONG LOT
HAYFIELD
WOOD LOT
REYNOLDS ROAD
LITTLE MOHAWK ROAD
PATTEN ROA

ADVENTURES IN YARN FARMING

Adventures in YARN FARMING

FOUR SEASONS ON A NEW ENGLAND FIBER FARM

BARBARA PARRY

Photography by Ben Barnhart

ROOST BOOKS
Boston & London
2013

Roost Books
An imprint of Shambhala Publications, Inc.
Horticultural Hall
300 Massachusetts Avenue
Boston, Massachusetts 02115
roostbooks.com

Parts of this book were first published in Twist Collective (www.twistcollective.com) in the 2009 series titled "Yarn Farm: Four Seasons from Sheep to Skein." Parts of this series are reprinted with their permission.

9 8 7 6 5 4 3 2 1

First Edition
Printed in China

♾ This edition is printed on acid-free paper that meets the American National Standards Institute Z39.48 Standard.
♻ Shambhala makes every effort to print on recycled paper.
For more information please visit www.shambhala.com.

Distributed in the United States by Random House, Inc.,
and in Canada by Random House of Canada Ltd

Designed by Daniel Urban-Brown

LIBRARY OF CONGRESS CATALOGING-IN-PUBLICATION DATA

Parry, Barbara.
Adventures in yarn farming: four seasons on a New England fiber farm /
Barbara Parry.—First edition.
Pages cm
ISBN 978-1-59030-823-3 (hardcover: alk. paper)
1. Sheep farming—Massachusetts—Anecdotes. 2. Wool.
3. Farm life—Massachusetts—Anecdotes. 4. Parry, Barbara. I. Title.
SF375.4.M4P37 2013
636.3'145—dc23
2013000373

For Mike and Caleb

CONTENTS

Summer 85

Fall 157

CASTING ON

A LIFELONG PASSION for working with animals, being outdoors, and creating things by hand led me to our farm. I had no template for building a flock of sheep, no agricultural background, no business plan. I suppose you could say my original starter flock was a bit like knitting a test swatch without a pattern. I simply picked up the needles and cast on.

While each season has its own tempo, there is no such thing as a typical day on my sheep farm. At five A.M. on a March morning I may find myself on my knees in my pj's in the small pen in the corner of the barn helping a ewe who has just delivered her first lambs. Before I've brushed my teeth or had my own coffee, I've helped her twins find breakfast.

Ten P.M. one November evening finds me in the pasture, having been roused from bed by the raucous yip of a coyote pack that has breached the fence surrounding my lambs. Working in the field by the headlights of my car, I patch the fence and make sure all the lambs are accounted for.

As a shepherd, I do some of my best work in my pajamas.

For some, raising sheep may seem like taking an awfully long and complicated way around the process of making sweaters. With the ready availability and staggering number of commercially produced yarns, why bother? I suppose it's for reasons similar to why I grow my own heirloom tomatoes and enough brussels sprouts for half of Franklin County, Massachusetts, or weave table runners and knit scarves for Christmas gifts or make bluebird houses from pine sustainably grown and milled at the tree farm up the road. Taking time to figure out how to make something from the materials on hand sharpens appreciation for the end product. Learning from others along the way enriches the process. Raising sheep has taught me so much about the intangible gains of and the best reasons for taking the long road. At heart I have a passion for taking an awfully long and complicated way around most things, drawn by the curiosity of blind journeys, the intensity of challenge, and a desire to see things come full circle.

My apprenticeship as a shepherd and my fascination with the process of making yarn from the sheep I have raised evolved from a midlife detour. In 1997, when winding down from an eleven-year stint teaching English at a small private school, I was ready for new challenges. A recently acquired taste for hand spinning was rapidly followed by a love affair with weaving. Both kindled a fiber addiction. I briefly flirted with the idea of pursing an MFA in textile arts at the University of Massachusetts—instead, I eventually enrolled in a master weavers program closer to home at the Hill Institute in Florence, Massachusetts. My flock then consisted of two sheep. I kept them in a small paddock in our backyard on a cul-de-sac in a rural subdivision in South Deerfield. Although I didn't know it at the time, my tiny flock and formative experiences with wool, lanolin, spindle, and loom were a portal.

My husband, Mike, who was then at the helm of Yankee Candle, the country's largest candle maker, was looking forward to a retirement of travel and rounds of golf. His work kept him running in fifth gear and away from home a fair amount of the time. When we were both burned out and ready for a sea change, the Berkshire foothills of western Massachusetts beckoned. Neither of us had a clue about what we were getting into when, in 2000, in what seemed like a capricious move to those who knew us well, we traded our home in the subdivision for a rambling 220-acre farm in the town of Shelburne. We packed up our son, the two sheep, and our three golden retrievers and headed for the hills.

Now our flock fluctuates in size anywhere from sixty to one hundred sheep (given the year and season). Each spring we shear sheep and keep vigil for lamb birthings. I take our yarns to the spring fiber festivals. In summer we set up temporary fences for grazing and maintain the fields. While our flock grazes, we work at maintaining the land: beating back the brushy-scrubby areas, clipping and hand pulling noxious weeds. We tend the garden. When the season hits the sweet spot of tall grass and dry weather, with our neighbor's help we put up hay for winter. Fall is the time for dyeing yarn, fiber festivals, and turning the rams in with our ewe flock. It's also time for the lambs we raise for the table to take their leave. We see our flock through winters of mind-numbing cold, snowdrifts up to the doorknob, wind strong enough to knock you off your feet, power outages, frozen buckets. While never "routine," our life here has come to feel quite normal to us.

Mike, once accustomed to the high stakes of meeting Wall Street's expectations, now enjoys brush hogging fields, bottle-feeding newborn lambs, and mucking out pens (even though it's not *exactly* how he had envisioned retirement). I love the trusting connectedness that grows between me and my ewes when I see them through lambing and how that connection deepens in me when I spin and dye their wool. I also love making connections with neighbors who stop to pitch in at the farm, with people I see year after year at the fiber festivals, and with our loyal members who subscribe to our yarn CSA (Community Supported Agriculture), some of whom I know I'll likely never meet because they live in another state or on the other side of the world. My work is enormously gratifying on so many levels.

This book recounts what I've learned about raising sheep, working with fiber, farming, and life in rural New England. The projects evolved as a collaborative effort with friends and neighbors. Writing this book has taught me, more than anything else, how much I still have to learn about this particular place and way of life I've come to love deeply. The more I learn, the more I realize I've only scratched the surface. I hope this book presents yarn farming as the closely interwoven fabric that it is: keeping sheep and working closely with land, resources at hand, and community.

THE ROAD TO SPRINGDELLE FARM

MY FIRST VIEW of Springdelle was a bleak drive-by in midwinter 2000. Mike and I received a tip from a local real estate agent who knew we were looking for a few acres and a barn for two sheep. There was a farm for sale on Patten Hill in the west-county town of Shelburne. I was already familiar with the Patten District. You know you've found it when, heading a few miles north of the Mohawk Trail, you come upon a very large red sign with a hand-lettered map. It lists most residents of the district and, though the map is not to scale, shows the approximate locations of their homes. Known for its sweeping seventy-mile views and as the location of the High Ledges, a six-hundred-acre Audubon sanctuary, "the Patten" is also home to a number of working farms: three dairies, a sugar maple farm, and a grass-fed-beef farm. The actively managed agricultural land preserves the vistas. Once you've reach the big red sign, the sky gets big, too, and Patten Hill reveals itself. Unless you visit on a leaden winter's day.

I had to rely on the diagram provided by my real estate agent because, from the roadside, there is no other way to get a sense of Springdelle's crooked footprint. The farm straddles the top and then zigzags down the eastern face of Patten Hill, an elevation change of roughly six hundred feet. The easternmost mowing, a large swath of hayfield at the farm's lowest elevation, peters out at a large beaver pond.

As we approached, we saw at the roadside a hulking dairy barn with a lop-sided rusty silo and a complex of well-weathered outbuildings that held the middle ground. A large pole barn housed farm equipment, a modest ranch housed the farmers. On the west side of the road the farm sloped radically upward in a patchwork of mowings, pasture, sugar bush, and pinewood criss-crossed by brooks and hemmed by stone walls. We had our hearts set on building a home, but between rock and slope, the landscape at first glimpse didn't appear to offer exactly what we had in mind for a homesite—until I spied a clearing at the very top of the tallest pasture. Looking uphill, I wondered, how the heck do I get there from here?

As it turned out, in the dead of winter the upper farm was accessible only by cross-country ski or snowmobile or by taking a two-mile detour via the town road (and to this day there's no true road through the farm itself, just a series of trails, almost always impassable in winter).

Days later, I returned for a closer inspection with the real estate agent. Mightily blanketed in snow, the lay of the land was still hard to read. I knew the soils on the hill were "thin," but it was hard to say just how much rock there was and how close it was to the surface. It would make a big difference in pounding fence posts and excavating a home foundation. Small divots in the snowy hillside hinted at the presence of springs right at the surface, but exactly how wet were the fields? And what was the quality of hay and forage? Outcroppings of barberry and juniper scrub told the tale that not all the land was intensively grazed.

The buildings were simple, functional, and loaded with signs of hard work. And cows. I looked skeptically at the roof of the big barn, capped by a half-shingled, half-patched-with-aluminum-sheathing roof. My tour of the working dairy at milking time was pungent and loud. I began to have misgivings. Stanchioned black-and-white Holsteins lined the aisle, each waiting her turn to be hitched to the vacuum hose that slurped the milk off to the enormous stainless steel tank in the milk room. The cows pointed away from the main aisle, each with her own helping of silage and individual automatic watering bowl.

The farmer proudly gestured to a large gutter about ten inches deep and fourteen inches wide in the concrete floor. It ran in a giant loop up and back down the center aisle just behind the cows. I nodded, though its purpose was not clear to me. He flicked a wall switch activating a chain to which a series of scuppers was attached. With a horrid mechanized din, the scuppers scooped their way through the channel, around the back ends of the cows, carrying a rancid slurry all the way to the northeast corner of the building, where it was flung out via an opening in the wall. "A gutter cleaner," the real estate agent explained.

Ahh.

Across the road was "a free-style" barn. That's what the farmer called it. Wondering if it meant the cows could do as they pleased, I mentally added it to the growing list of things I saw that didn't make any sense to me. (I later learned it is actually a "free-stall" barn because the heifers *can* come and go as they please.) The heifers stood shoulder to shoulder at the feed bunk that ran down the entire south face of the barn. Behind the building the snow was pocked with cow pies.

If at the time I had had any true understanding of the workings of a dairy, I could have more intelligently appreciated the efficient layout of the buildings: how the big barn was built into the slope and designed so that all the work flowed downhill. On the uphill side, closest to the road, was a large sliding door through which loads of sawdust could be dumped directly into a cavernous stone pit one floor below—conveniently adjacent to where it was needed to sop up the mess in the dairy. The hayloft was accessed by a high-drive door, where wagons could be backed right into the barn at loft level for ease of unloading. The loft itself was monstrously vast. Trapdoors at intervals in the floor allowed for bales to be dropped at strategic points alongside the feed trough in the dairy below. And the manure shed, which I couldn't bring myself to approach, abutted the downhill side of the big barn—where the gutter cleaner spewed its contents.

There was no doubt about it, it was the real deal. While not exactly picturesque in the iconic New England postcard sense, I found the place both overwhelmingly authentic and peculiarly charming. I had no idea what we'd do with all of that land, space, and equipment. And I never thought Mike would go for it. Part of me was sold before I ever saw the upper farm and the view from the clearing at the hilltop. Although my common sense said, "You've got to be kidding!" my intuition told me it was a rare opportunity. I began to consider the possibilities.

Within a week, Mike and I saw the view from the top. The clearing I had spotted earlier from down below was reached via a cart path that led from the town road two miles from the doors of the barns. On a clear, bright afternoon we hiked in following a compacted trail made by snowmobile traffic. The trail wended through a hemlock stand, then opened up into a birch-ringed pasture. It curved through what might once have been an orchard; just a few gnarled apple trees remained. Nearly half a mile in, the trail hooked right at a stone wall that had a cow skull resting on top. And then the woods gave way to sky.

Before us a large pasture undulated and unfolded, flecked with islands of birch, wild cherry, and poplar wherever the bedrock poked through. Beside a clump of hardwoods was a fairly level plateau with an arresting east-northeast view that stopped us in our tracks. To the northeast, Mount Monadnock stood against the blue sky, crowned in white; directly east were Greenfield's chief landmarks and, beyond that, the Connecticut River watershed and the Quabbin hills. The pasture was edged with woodland of white pine, birch, and mixed hardwoods. From the clearing we could just make out the roofs of the barns below. In a move uncharacteristically impulsive for both of us, we quickly made an offer on Springdelle Farm.

A decade later, our flock calls the barns that once housed Holsteins their home. Although the basic footprint of the farm compound is unchanged, the interior configurations of all the various buildings have been modified for keeping sheep. Our rams winter in the building that was once the machine shed. The free-stall barn, minus the cattle stalls and gates, protects the ewes in winter. Before leaving, the farmers had at our request dismantled the milk pipeline and removed the gutter cleaner and the eight-hundred-gallon bulk tank from the milk room. We have chopped the steel stanchions from the concrete floor of the dairy and installed fifty feet of feed bunk for sheep. Lambing pens made from rough-cut pine now line its east wall.

Over the years, we've snipped and balled miles of barbed wire fencing—which proved dangerous to woolly sheep—replacing it with high-tensile woven-wire perimeter fences around many of the pastures.

We've made our home in the clearing surrounded by white pine and birch at the top of the hill, overlooking the barns below. In summer, the sheep graze the pasture just steps from our back door. The rough track that led from the town road to the pasture with the big view is now our driveway. To this day, unless we feel like taking a long walk through the woods and fields, it's a two-mile commute from our house at the top of the hill to our farm even though they are on the same property.

FLOCKING

IT IS A UNIVERSAL TRUTH that a shepherd in possession of sheep will be in want of more sheep. When we moved to Springdelle, I wanted more sheep in a hurry if for no other reason than to avoid feeling ridiculous every time one of my new neighbors inquired about the size of my flock. I had scads of barns and miles of fence and field. With unbridled enthusiasm and visions of being an übershepherd, I couldn't wait to get going.

My first ewes, Cocoa and Chablis, were a genetic potpourri. They were mostly Border Leicester in body style: open face (clean of wool), arched Roman nose, prominent brow, wide bug eyes. Their Leicester genes imparted length to their fleeces, which reflected their Rambouillet ancestry in fineness and crimp. When Cocoa delivered triplets in her third year and quadruplets a year later, I learned from Jean Willmann, our neighbor from whose flock Cocoa and Chablis originated, that this was not unusual since there was also a hint of Finnish Landrace in their blood. Finnsheep are known for being prolific.

To learn more about what the world of wool had to offer, I seriously scoped out the fleece breeds at the sheep and wool festivals that year. Although all sheep grow wool, so much of the American sheep industry is focused on raising market lambs that wool is regarded as an afterthought with little value. Dorper sheep, raised strictly for meat, actually shed their fleece, eliminating the expense and chore of shearing. In the show ring, the sheep are sheared slick because the judges are interested in the carcass, not the fleece.

The best place for comparing sheep side by side is in the breed barns at sheep fairs. I made my notes of the major wool breeds. Merino is best known for having the finest wool and a high yield; they were tempting, but as a rookie shepherd I was intimidated by their spiraling horns. The Rambouillet, essentially Merinos illicitly smuggled to France, are larger framed and looked like they'd be a lot of sheep for a five-foot two-inch shepherd to wrangle. I wanted a flock that I could easily manage for the basic tasks of deworming and hoof trimming. I considered Shetlands, an ancient breed with fine-fiber wool (traditional shawls knit from yarns hand spun from their neck wool are gossamer webs, fine enough to be pulled through a wedding ring and thus called "wedding ring shawls"). They are diminutive, and though the rams come with horns, I was tempted by their tantalizing range of natural colors. Jean explained that they were a primitive "unimproved" breed and therefore grow two distinct types of fiber within the same fleece: a fine, shorter wool for insulation interspersed with coarser, longer fibers to repel the elements. This factor requires an extra step of separating the two fibers before spinning. I've since learned that many Shetlands are bred to have more uniformity, eliminating the need for this.

On the opposite side of the wool spectrum were the longwool breeds: Lincoln, Border Leicester, Leicester Longwools, Bluefaced Leicester, Cotswold, Coopworth, Wensleydale, Romney, and Teeswater. Attractive to hand spinners for the length of staple, the word used to describe a lock of wool—a year's growth ranges anywhere from six to ten inches—the fleeces of several longwool breeds also have an amazing luster.

To further complicate matters, there is a host of breeds classified as "medium wool": Finnsheep, Tunis, Corriedale, and Montadale.

Whenever I met breeders at the shows, I inquired about their sheep. All were eager to list the pros about their breed, especially to a prospective buyer. And that's when I learned about pride and prejudice and sheep. Shepherds are passionate. Those who show in breed competitions take it seriously and make no bones about sharing what sets their sheep apart. I observed this while walking the show barns and reading each farm's sign beside its pen: CORRIEDALE SHEEP—THE BETTER BREED. BEST TUNIS. It was a crash course in Sheep Breeds 101. I met some wonderful folks who were very helpful and generous with their time. Mike patiently endured hours of sheep talk, both at shows and at home in evenings when I chatted with breeders over the phone.

I fell in love with nearly every breed I saw that year: Jacobs for their bizarre multiple sets of horns and wild dalmation-like spots, Shetlands and Icelandics for their rich color and unvarnished beauty, sporty-looking Cheviots, adorable Babydoll Southdowns. Feeling pulled in so many directions at once and sorely tempted to acquire sheep in the same manner in which I once shopped for shoes (a pair in every style), I needed some unbiased guidance. For that I turned to someone who was intimately familiar with both sheep and wool—my shearer, Andy Rice.

Andy was prompted to learn sheep shearing thirty-five years ago. His wife, Linda, kept a couple of ewes for hand-spinning fleeces. The person Linda hired one spring to shear her ewes nicked them up so badly in the process that a friend asked Andy if he had tried to shear them himself. Andy's testy reply was, "No, I *paid* someone to do that to them." And he resolved to learn to shear sheep himself.

He apprenticed with shearer Bruce Clement and also attended eight New Zealand shearing schools in the United States, working directly with noted New Zealand wool board shearing instructors Alan Barker and Rex Kringle. As his shearing vocation evolved, Andy also worked as an ambulance driver and an EMT. His emergency medical skills with people eventually came into play in his work with livestock. Clients, many new to the world of raising sheep, came to him with questions about sheep husbandry and, especially, with sheep crises. Calm in the face of emergency, he demonstrated a knack for sorting out a crisis, often over the phone, figuring out who really needed help: the sheep or a nervous shepherd.

In addition to shearing, consulting evolved into a business, though Andy still shears about 2,500 head of sheep, along with goats and llamas, in the course of the year. Over time I came to rely on his expertise as both a careful shearer and a good person to turn to for help.

Although Andy is today a trusted friend, at the time I was initially shy about asking for advice from someone I barely knew. Our rapport to this point was limited to his annual visits to my backyard flock on shearing day. But after bumping into Andy at the Massachusetts Sheep and Woolcraft Fair and telling him about my new farm, he told me that in addition to shearing, he was available for flock management services and consultation. He'd be happy to look at my new farm's setup and to give me some pointers on building a flock.

As we toured my barns and pastures one afternoon, I broached the subject of choosing sheep. He was quiet for a moment, and I was afraid I'd put him on the spot. Was it awkward to ask him to speak candidly about sheep, given his clientele? He adroitly responded with a few questions of his own: "What are your goals? Do you want hand-spinning fleeces? How much time do you want to spend on maintenance like trimming hooves?" Noting the farm's pockets of rough terrain overrun with brush and bramble, he suggested I consider one of the longwool breeds such as Romney or Border Leicester—thrifty sheep that could be pressed to clean up my pastures that were reverting to scrub and still produce healthy lambs and fleece. I was already enamored with Border Leicesters, their fleeces having caught my eye—robes of long, shining ringlets. I also loved their elegant heads and regal carriage. Longwools were definitely my style. Another attractive characteristic of longwools is the low grease factor. Grease, also known as lanolin, is the waxy substance produced by glands that makes a sheep's fleece water-resistant. The lower the grease factor, the higher the yield of clean wool after washing. The low grease factor also meant it would be easier to hand scour the fleeces myself. He pointed out that they are easy to work with and not too large for a person my size to handle.

Finding the right sheep breeder depends on word of mouth. Through the grapevine I learned that a reputable breeder (who also trained border collies—a detail I filed away for future reference) lived in Lincoln, Massachusetts, just minutes from Walden Pond, in the neighborhood of the Gropius House and the DeCordova Museum. I was somewhat surprised to learn of a sheep farm in the suburbs of Boston, where real estate is at a premium. That summer I visited Betty Levin to meet her Leicester flock.

Betty came down from the field where she had been working her collies on some sheep, shepherd crook in hand, to greet me. The shed was dark. A group of small Border Leicester ewe lambs were penned inside, awaiting my arrival. I was instantly drawn to a very friendly little white one with a slightly runny butt. She approached me while the rest of the group retreated to the corner. "Oh, I don't recommend that one," Betty quickly said. Reminding myself that I was a sheep farmer looking for breeding stock, which is not like choosing a puppy, I closely studied the others.

One black ewe stood out for having a sculpted head and noble demeanor. Though she was young, her fleece already had nice length, starting well behind her erect ears and fanning out like a ruffled Elizabethan collar. She was very "typey," displaying an exaggeration of the breed's chief characteristics. Another, smaller, ewe had a stunning dark fleece but more rounded facial features. "One-eighth Black Welsh Mountain," Betty explained. A third caught my eye. She had a dark fleece with tips lightly peppered in auburn. All appeared straight legged and clean. Satisfied with my choices, I shook hands with Betty. Later that summer I returned to her farm with my trailer to pick up my Border Leicester ewes: Elizabeth R., Jenny, and Annie.

No matter the size of a flock, it is standard health protocol to place newcomers in a quarantine pen before introducing them to the rest of the flock. I set the new Border Leicesters apart from Cocoa and Chablis, dividing the space in the free-stall barn with the inherited cattle gates. It seemed like a good idea. My new ewe lambs turned out to be gate-crashers (perhaps *gate scoochers* would better describe them). Unsettled by the two-hour trailer ride and understandably unnerved by separation from their native flock, after I left the barn for the day they did what sheep do—they flocked up with the only other sheep they could see at the far end of the barn. They were small enough to scoot beneath the cattle gates. They had not read the memo on good health protocol. It was my first in a series of earnest but ultimately futile efforts to make the farm's existing cattle setup work for sheep.

Border Leicester sheep proved to be easy keepers. I tended to feed them heavy-handedly, so my little ewes became tubby bodied, but their legs remained spindly. Funny looking—but otherwise hale and hearty. As my flock evolved, so did my tastes in fiber. As a spinner and a weaver, I loved working with the Leicester fiber. I could spin perfectly smooth yarns strong enough for warping looms, durable enough for socks. As I began testing the waters of marketing fleece and yarn under my newly launched business label Foxfire Fiber & Designs, I listened carefully to customer comments. Spinners sought Leicester fiber and loved its durability and sheen. But I often heard knitters comment that my yarns, though lovely, felt rustic. I sometimes heard whispers of "scratchy" when they held a skein to their neck. True, my yarns were smooth, dense, and lustrous, but there was also a prickly factor. Even the finest, silkiest Leicester fleece cannot compete with Merino for softness. But then a sweater or a pair of socks knit from Merino fiber would not hold up to wear the way Leicester would.

To satisfy my own shifting fiber preferences and to remain competitive in the wool market, I investigated other breeds. Although Merino was the most logical counterpoint to Leicester, I was still somewhat daunted by the thought of owning a horned ram. My quest for fine fleece led to my first encounter of the Cormo kind—at Alice Field's Fox Hill Farm in Lee, Massachusetts.

In the world of sheep, Cormo is a relatively new breed developed by Ian Downie in Tasmania in the 1960s in a quest to build a better wool producer. Technically speaking, the genetics behind the original breeding were one-quarter Lincoln, one-quarter Australian Merino, one-half superfine Saxon Merino achieved by crossing Corriedale rams with a select group of superfine Saxon Merino ewes. Downie pioneered a breed nearly as fine fleeced as Merino (a micron range of seventeen to twenty-three according to the breed standards of the American Cormo Sheep Association). Offspring were selected for high-yielding fleeces, good growth rate, and high fertility. They were introduced to the United States in 1976 and over the past decade have gained favor among shepherds wishing to improve fiber quality.

Compared with my angular and statuesque Leicesters, Alice's Cormos reminded me of round woolly teddy bears. Where the Leicesters have angular, bony heads, the Cormos' heads were poufy with topknots like dandelions gone to seed. Unlike the Leicesters' clean legs, the Cormos' legs are covered in wool. Alice kept their white marshmallow-like fleeces pristine by dressing her flock in special sheep coats. Unlike Border Leicester fleece, which grows in open locks, the Cormo fleece is dense enough for dirt, hay chaff, and hitchhiking seed heads to become embedded in the fleece if left unprotected. The stickiness of the wool caused by the high lanolin content makes it extrahard for debris to be removed in processing. Clearly Alice was a stickler in maintaining her sheep and fleece, and I was immensely impressed.

Alice and I sat down at her kitchen table to talk about Cormo fiber. She brought out a box of little ziplock plastic bags, sorted by year, each holding a lock of wool. Like time capsules, there was one for every sheep, a sample taken on shearing day to represent that sheep in that moment of time. The locks of wool spread out on her kitchen table told the story of the lineage of three ewes I wanted to purchase, Buttercup, Pansy, and Charlotte, all descendants of Mario, one of her main herd sires. I had met Mario, a perfect gentleman, during my tour. Then she showed me the breeding behind Trumpet, a somewhat feisty

but promising ram lamb I was purchasing from her to cross with the ewes. I was fascinated to see the bearing that a ram had on a breeding program and how it could alter the course of the program. Alice's passion for raising healthy sheep and quality fleece was clear, and I was both impressed and inspired by her work. My first visit to her farm was a revelation and also one of my favorite afternoons ever spent in the company of a fellow shepherd.

We welcomed six of our own Cormo lambs to the barn the following spring. Our Cormo ewes shared the feed bunk in the birthing barn with our Leicester ewes while our mixed brood of lambs bounced from one end of the barn to the other. Our flock continues to be a bit of a hybrid. Cormo fine wools now outnumber the Border Leicester longwools. A percentage of ewes are direct descendants of Cocoa. She is our original matriarch. Her daughters have been bred out to Cormo rams, which greatly improved the style and consistency of our wool clip. Eventually, we've added Merino-cross Moorits (from Alice) to breed for brown, a recessive color gene in sheep. More recently we have introduced Shetlands for my own hand-spinning fleece. Llamas joined the flock;

they serve the role of lamb nannies and bodyguards. And for raising mohair and trouble, we've added a handful of Angora goats. Our yarns today reflect our flock's diversity: Cormo blended with mohair; fine-wool crossbred wool carded with alpaca fiber from a nearby farm. As with any form of agriculture, the earth and sky ultimately call the shots. We are committed to raising healthy sheep while striking a balance with the number of head our land can support.

In our second year at Springdelle we tried our hand at raising beef, acquiring a pair of white Galloway heifers in late autumn 2002. We kept them for about five months, spanning a cold and snowy winter, which was long enough for us to discover that when it came to cows, we were way out of our comfort zone. Josephine and Fidelia were a fractious pair. Having had free range of a large pasture and little human contact in their former home, they were nearly feral in temperament. Working around them in the confines of the winter barn was an unsettling experience. Josie would kick me if I moved too close to her. They roughed up the goats and harried the sheep. Their frozen cow pies were like mounds of cement. After assisting Josie with the delivery of a bull calf that April (or, I should say, watching my neighbors John and Carolyn help Josie deliver her calf—no easy task since it was difficult to get near her), I decided to call it quits with cows. Fidelia, Josie, and the bull calf walked up the road with John and Carolyn to join the herd next door at Wheel-View Farm.

To keep our flock safe from predators in 2003 we added a llama, Crackerjack, to serve as sheep guardian. Llamas' innate behavior to approach and investigate a potential threat makes them good protectors for sheep, whose instinctive defense is to flee and huddle. Crackerjack proved so useful and reliable, we eventually added two more llamas, Caitlyn and Sol, to our flock.

MAX0104

SPRING

The hayloft is the last place I want to be on a March evening when winter is delivering its final blows. Although I have a very sensible rule about making sure enough bales are brought down each morning, on this occasion I had neglected to toss down enough bales for dinner for the sheep. In the darkened loft the rafters creak. The barn sways as the north wind strikes the gable end, sifting a fine powder of snow through the chinks and knotholes. I feel my way through the shadowy maze of stacked bales, over to the hay chute on the west wall.

Fumbling for strings with gloved hands, I yank four bales from the top of the stack. I slide the bales across the smooth wooden planks toward the plywood trapdoor in the floor. Lifting the door, I shove an evening's ration down the chute, listening as the bales hit the concrete floor below. There's a wild, impatient rustling from the ewes penned in the straw beneath me. At last, dinner is served.

This is the tipping point of the year. In winter the farm pauses and holds its breath. Now, after frozen months of sitting still, the farm comes out of hibernation. Spring feels like the start of a new journey. The barn is my vessel; we set a course for a new year. Shearing, skirting fleeces, and lambing are now visible points on the horizon. The yearly rituals are the same, but no two years are alike. We enter the paradox of the familiar and the unknown.

GOOD FIBRATIONS

WHEN WINTER eases its grip, the icy back roads of western Massachusetts thaw from glacial hardpan to mud pudding. I estimate about a third of Shelburne's town roads are gravel, not counting the roads down in the village of Shelburne Falls. Nearly all, including my own, are a complete disaster in mud season. A simple trip from the farm to the mailbox can be a white-knuckle adventure. As I round the bend near the orchard, I drop the car into low. Mud sucks at the tires. Ruts fight me for control of the wheel.

Sawhorse barricades appear all over town. ROAD CLOSED. PASS AT YOUR OWN RISK. MUD IN ROAD. With all of my back-road shortcuts to the village closed, I make the trip to town the long way around, sticking to blacktop.

As my neighbors ready their buckets and sap lines for the spring sugaring, I prepare for my annual wool clip. The ewes are marshmallows, burgeoning with a year's worth of wool on their backs and unborn lambs in their bellies. They've waddled a well-worn path that runs out the door into the sunny south yard behind the barn. A soft *whoosh* announces their comings and goings as they brush against the door frame.

Shearing day has been circled in red on my calendar since I pulled the ram from the breeding flock last November. When our lambs are due to arrive in March, the "spring" shear falls technically in late winter. Shearing at this time of year, four weeks before lambing, is part of the seasonal clockwork at many sheep farms. It's a practice I adopted after several years of shearing much later in the spring, when the lambs and flies are out and a year's worth of precious wool is grimy and sticky from lambing. A young lamb's favorite climbing apparatus and napping spot is its mother's backside. In a year when shearing followed lambing, I pitched several otherwise fine fleeces completely trashed by the pitter-patter of tiny hooves. Rather than struggle to teach a newborn lamb the difference between momma's teats and momma's manure-

tagged wool, I changed my management. Shearing before lambing now avoids all of that.

I also have my own selfish reasons for harvesting the fleeces prior to lambing. I learned early on that nothing is more painful than discovering fleece flaws on shearing day. Even with the best pre- and postnatal diet (my girls get high-quality second-cutting hay supplemented with grain), the stress of lambing and nursing can cause tenderness in a ewe's fleece. Shearing in late winter places that potential weak spot right at the transition point between old and new fleece rather than at some point midstaple. In preparation for shearing, Holly, my barn assistant, and I plan and strategize as we sip our morning lattes. Which group goes first? How shall we arrange the pens? Can we keep everyone dry under one roof if

it's raining? Do we have enough clean sheep coats to clothe the entire flock? We make sketches indicating pen configurations and sheep traffic flow.

When my flock was much smaller, my shearer, Andy Rice, and I used to manage shearing alone. In my years of raising longwools, before I took a liking to Cormos, we could easily blow through twenty sheep in a few hours. As with most fine-wool sheep, a Cormo takes longer to shear. The fleece is dense and sticky. The breed has inherited skin folds that act like speed bumps and emulate the easily nicked thin skin of its Merino cousins. My flock is now too large to shear all in one day. Planning for shearing is now a careful orchestration.

On the day before shearing, Holly and I sort and organize the sheep wardrobe. Colored bits of fabric sewn into the chest seam on the front indicate sheep coat size. Sheep aren't the least bit self-conscious about wearing their dress size on their chest for all to see. The most matronly ewes are in silvers. A few smaller ewes are in tans. Once sheared, everyone drops one coat size. We figure out in advance how many we need of each size.

Early on the morning of shearing day, before Andy's arrival, we gather the flock from the various outbuildings. Gathering boys and girls in close proximity creates a stir that makes it difficult to maintain a mellow vibe. Fortunately, the temperature is above freezing and there is little danger of sheep skating out of control as we lead them down the incline to the shearing barn. The ewes are curious. Although they've not set foot in this building since last summer, they instinctively walk right in. Next come the rams and wethers. The boys' interest is clearly piqued by the presence of ewes, and even though none of the ewes could possibly be in estrus, we keep some buffer space between the pens.

One by one, my shearing team assembles. This year's group is made up of trusted, dedicated friends who hail from different walks of life but enjoy the occasional shearing gig. Gale, who works part-time as a veterinary technician, helps Holly arrange the sheep coats. Laurie, a local artist, arranges the items she needs for bagging and labeling wool, while Susan, a literary agent who is new to the team, gets the lay of the land. With help from Caleb, a part-time woodworker and sailor, Mike tightens up the holding pens. The sheep stand in close huddles, which puts them at ease. They begin to let their guard down. Unlike people who get claustrophobic, sheep feel safer in crowds.

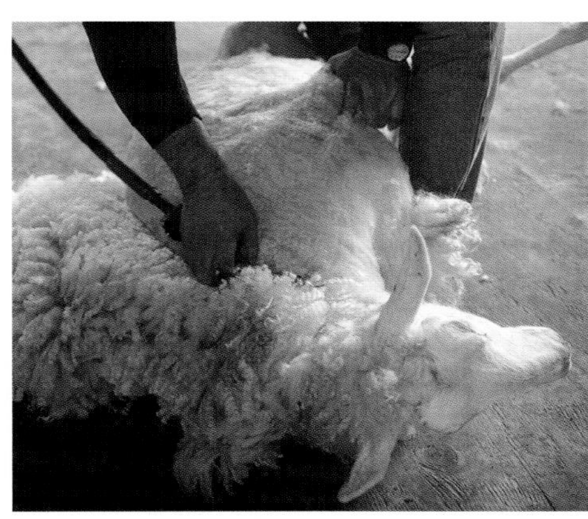

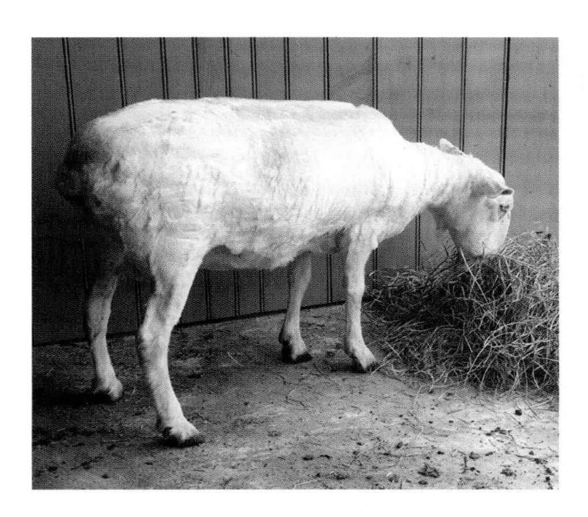

Andy arrives. While he assembles his hand piece—the electric shearing blade and comb he will use on my Cormo flock—we usher the sheep into smaller pens to make handling easier and begin stripping away their jackets, which have protected their fleeces from dirt and chaff all winter long. The coats are filthy; the fleeces are not.

In the years before my flock had its own wardrobe, shearing-day results were a mixed bag. The fleeces were gloriously long and lovely but also heavily seasoned with dirt, seed heads, flakes of grass, and bits of stem. Wool tips were faded and weakened by exposure to the elements. So much otherwise perfectly fine wool ended up on the compost pile.

Sheep coats became compulsory about six years ago, especially during the long months of winter confinement. Most of my flock has been wearing coats all of their lives and know the drill. They stand like toddlers as we lift hind hooves one at a time and slip off the leg straps. We wriggle the elastic free from the deep channels of wool and skin folds at the neckline and carefully work the coat up and over their woolly topknots, taking care not to snag an ear. The sheep stand patiently as we unveil and admire each fleece.

Mike draws Cassandra, the first ewe from the pen, onto the shearing board. Andy pulls the cord that starts the purr of the electric shears; their background hum becomes the mantra for the day. Turning Cassandra's head back over her shoulder, he gently tilts her onto the board. Balanced on her rump and cradled against the shearer's legs, Cassie senses there is little point in struggling from this position and comfortably settles into a Zen-like trance. Andy begins tracing the shearing pattern with his blade.

The process of separating sheep from fleece is a bit like unzipping a baby in a bunting. Andy starts by unbuttoning the belly wool with a series of short strokes of the shears called "blows." Protecting the udder with his hand, he works the blade carefully around the crutch, the ewe's hindquarters. He next unfleeces the left rear leg, then unzips the upper portion of the fleece at the inside of the neck like a sweater, by working the comb upward from brisket to chin. He cleans the face and strips the left front shoulder. A deft 90-degree pivot of a ewe on her fanny is followed by the long blows that run the entire length of the sheep from tail to ears. He then strips the right flank. There is no rushing here, this is not a race. Working entirely within the moment to the rattle and hum of the shears, we breathe and channel our collective energy. The sheep stay mellow. The fleeces are exquisite.

Each of us has a role, and we work together in rhythmic choreography. Mike patiently extracts a reluctant ewe from the holding pen. As Andy shears, I kneel on the shearing board, positioning myself across from him, where I have the best view of the fleece as it comes off Cassandra and her condition as she is disrobed. I perform a rough skirting as the fiber falls to the board, removing unwanted bits: pitching manure tags, tossing butt and belly wool into the junk bucket, and flicking away any second cuts—tiny shavings of wool created whenever the shearing blade has made a second pass in the same spot. I toss the "fribs"—neck wool that's unsuitable for yarn but may be salvaged for other uses—into another bin. I pause to cup the ewe's naked-of-wool belly, palpating her unborn lambs. I skirt, Caleb sweeps, and Mike readies the next ewe.

At last, Cassandra springs from the shearing board. Laurie shakes open a large plastic bag. We carefully roll the fleece into a fragrant bun that radiates warmth and the essence of good sheep karma. Snatching a staple from the bag, I pause to examine and assess. Cormo wool should be fine and uniform in crimp, snowy white in color. I hold the staple to my ear and, tugging at both ends, give it a snap and listen. A clean *ping* signifies a healthy fleece. *Snap, crackle, pop* means tender wool that spells p-i-l-l-y yarn.

Cassandra's fleece is sound, evenly fine, and glistening with lanolin. Nirvana.

The freshly naked ewe enters the wardrobe pen, where Holly, Susan, and Gale fit her into a new coat, a size smaller, and a lot cleaner than the one shed fifteen minutes earlier. She needs little coaxing back into the fold, where fresh water and a manger of sweet summer grass await.

The mound of wool bags climbs as we work our way through the flock. Holly tends to the freshly sheared ewes, who jostle and spar at first. It takes them a few minutes to adjust to an instant ten-pound weight loss and to recognize their own sisters.

When we break for lunch, the entire crew trundles over to the studio, where a hearty meal is prepared for us by my good friend Margaret, who also happens to be a great chef. Mucky barn boots get unlaced and left in the mudroom by the back door. Our hands sticky with lanolin, tufts of wool stuck like feathers to our sleeves and hats, and well perfumed with sheep aroma, we make our way from sink to table.

The "studio," a modest ranch-style house nestled into the south side of the barn compound, is quite literally my home away from home. The primary residence for the farmers who sold us the farm, it is now used mainly for our camp during lambing season, when it's important to be closer to our flock at all hours of the day and night. During the rest of the year the studio serves as my work space for sorting fleece and yarn. It houses spinning wheels and looms and functions as a retreat where I can focus on my own hand-spinning, weaving, and writing projects. I've outfitted the garage as my dye studio. Since fiber work is continual throughout the year for me, it is essential to have a space dedicated to this process—and extremely handy to have in close proximity to my flock.

Shearing lunch is one of my favorite moments in the day. It's also the bribe and reward for my friends who come to help—and so it has to be fabulous. Exhausted and famished, we inhale a platter stacked with sandwiches: smoked turkey with arugula, sundried tomato pesto, cucumbers, and avocado on olive ciabatta; roasted portabella with wilted spinach, roasted red peppers, caramelized onions, and chipotle aioli on baguette. Margaret has also supplied a hearty soup of curried sweet potato with ginger crème fraiche. Our meal is rounded off with a dessert of lemon bars with pistachio crust and honey oatmeal cookies with dried cranberries and sunflower seeds. Coffee and convivial conversation follow lunch. No one feels like moving. Conversation continues. Andy glances at his watch. We still have more than half a barnful of sheep left to shear and less than half a day's worth of hours in which to do it. Time to get back to work.

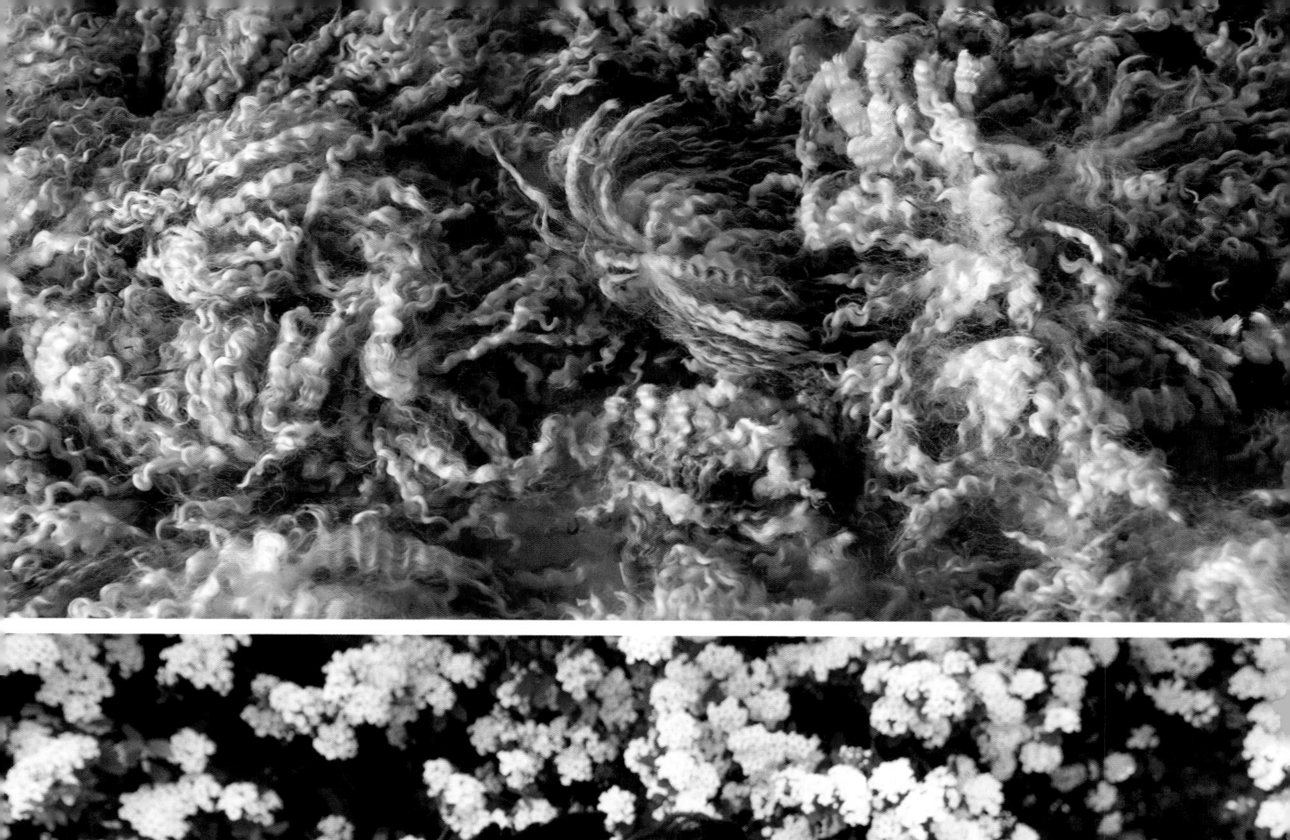

MASS.
SHEEP
AND
WOOLCRAFT
FAIR
FIRST
PLACE

THE IDES OF MARCH

IN WINTER'S WAKE, the farm's tempo radically changes. Suddenly there is much work to be done, in every corner of the barn and studio. As the snowbanks lining the roads continue to recede, we work our way through shearing the rest of the flock on successive days. With each round of shearing, the stack of wool bags in the wool shed climbs higher, like snowdrifts in the not-so-distant winter, only these white woolly drifts are indoors, stacked to the ceiling of the shed until there's barely room to turn around. Each bag represents at least another half hour's work of gleaning and inspection, which makes me feel panicky when I think of how much must be accomplished before lambing. I now feel that I am racing against the clock to skirt and box those fleeces and send them on their way to the mill before the crocuses sprout and the lambs emerge.

At my busiest times I arrive early at the lower farm to get started. Although the landscape is still stark and barren, vitality is in the air. The sheep have a case of the morning giddies. I sense an energetic buzz the moment I set foot in the barn. Unburdened from the weight of their heavy winter fleeces and enlivened by the shift of season, the yearlings prong leap like lambs from one end of the barn to the other. The ewes tussle in unladylike skirmishes. With ears straight back, Comet and Helena lock gazes; there's a moment of tension followed by the *boom* of a direct cranial crash. I tell them to knock it off as I shake a bale of hay open into the feeder. The girls settle their differences and get down to the important business of breakfast.

The breakfast barn reminds me of lunches in the dining hall at the small private school where I taught for eleven years. Crackerjack, the flock's guardian llama, presides. Like a ninth-grade student seated at the head of the table attempting to garner control of unruly underclassmen, he reminds them: napkins in lap, chew with mouths closed, and elbows off the table please. He politely waits for me to distribute the flakes down the line of feeders and for the sheep to stop tripping over one another before approaching an open space near the wooden pillar. Before he can make it, he's cut off by a pushy ewe. In quiet resignation, he shuffles over to another open spot. A moment later, he's shouldered again by a bossy yearling, who goes just a little too far. Resignation turns to indignation. Crackerjack leans over and glares in quiet warning, neck lowered, ears back. The gentle reprimand is all that's needed, and the yearling backs off. The perfect proctor, Cracker keeps his crew in line.

A pair of geriatric Cormo ewes, Charlotte and Buttercup, are sequestered in the pen adjacent to this rowdy crowd. They've caught the energetic vibe, too, tossing their heads in excitement as I fetch their morning grain pans. They meet me at the gate, gobble with gusto, and then optimistically search my outstretched hand for more. It's amazingly simple to make them happy.

In the studio, my day's work is cut out for me. Down in the basement near the washing machine, a reeking pile of grungy sheep coats awaits laundering. A waxy grime lies thick on the surface of each coat, a winter's accumulation of lanolin and dirt thick enough to scrape with my fingernail. Washing only four at a time in hot water is the only way to break the grease, and sometimes more than one washing is needed. I start another load. Yesterday's clean coats are spread out on the skirting table, where I had set them to dry the night before. While waiting for a pot of coffee to brew, I fold and neatly arrange them according to size. Like a stack of clean pinafores, back into the barn cupboard they go.

A skirting table is specially designed for laying out a fleece for inspection and sorting. Although I've seen many different designs constructed from a variety of materials, the basic elements of a skirting table are the same: a frame, either wood or steel, covered with a slatted surface (such as rows of PVC pipe with one inch spaces between them or with chicken wire) and resting on sawhorses. Bits of straw, manure tags, and second cuts fall through the slats or spaces in the table's surface as you work your way around the fleece. The surface should be large enough to spread a fleece so it sits completely on the table, with the cut ends facing down, wool tips facing up. The size of the table depends on the size of the fleeces. My smallest, a 4 × 4 foot wooden frame covered with fence mesh, works well for hogget fleeces. For ram fleeces I use a large, 8 × 5 foot frame that takes two people to set up. No fiber farm should be without a well-designed skirting table.

Tackling the sea of wool comes next. The bags lugged over in batches from the wool shed so I can skirt indoors in a heated space are literally everywhere. In the mudroom, in the kitchen, in the living room, lining the hallway. It's almost

impossible to turn around without tripping over a bag of wool. I'm afraid to let anyone inside the door at this time of year, lest they question my sanity. It looks really overwhelming because it *is*. I begin to wonder how I'll ever get all of this wool out the door and to the mill before lambing.

The answer is, one bag at a time.

So I start with a small ewe's fleece. The bag is marked 89, Calypso. Buttercup's daughter. The bold crimp and whiteness of this particular fleece tell me so without consulting my flock list. Wool traits are highly hereditary, which makes it easy to read a ewe's pedigree by examining a lock of wool. Calypso carries the mark of her mother's good genes.

With a shake and a *swoosh* the fleece slides out of the bag and lands like a dollop of whipped cream on the skirting table. With the cut side facing down, I carefully unroll and spread out the fleece, taking care to keep it intact. Once unrolled on the table, it resembles a topographic map of a sheep. I search the edges, looking for defining signs that mark the various regions of a sheep's anatomy.

Wool tipped in manure marks the posterior terrain, which is bordered by the urine-varnished wool of the breeches. I gently pull and separate the tagged and stained wool from the perimeter and toss it into the bucket under the table. Traveling up the flank, I identify patches of fine and kinky belly wool. Resembling silly string, this, too, gets tossed into the bucket under the table. The shoulder and neck region is marked by a deep channel in the fleece created by the elastic neckline of the sheep coat, a clear delineation of where the sheep coat begins. Everything "north" of this Mason-Dixon line is gray and full of chaff. To the south lies clean and protected wool destined for yarn.

I continue to work my way around the fleece, pulling any fibers that are soiled or cotted. The series of careful subtractions eventually leaves only the snowy "blanket," the white hills of the interior regions, on the table. The outline of heavily skirted blanket still resembles the shape of a sheep but reduced in size. And it definitely still smells of sheep.

With my hands well greased with lanolin, I grasp the edges of the blanket and shake vigorously to dislodge any tiny neps, or second cuts, made by the shearer's blade, not desirable but difficult to avoid when shearing fine-wool sheep. A dusting of fine wool bits sifts through the slats of the skirting table and onto the floor.

Before boxing this fleece, I snag a small sample lock from the shoulder region. It serves as a little report card, representing this ewe's work in wool production for the year. It goes into a small ziplock bag. With a pen, I record the name, ear tag number, and date on the bag. Like a mother saving a lock of hair from a child's haircut, each shearing I keep a sample lock of every sheep sheared on the farm. Boxes of fleece samples, sorted by year, get stored away in the cedar closet.

Grabbing another bag of fleece, I begin the process again. Though grimy, the task is one I rather enjoy. As I closely study each fleece, I see the genetic blueprint of my flock, which wool traits carry from ewe or ram to daughter. I mentally note my favorite fleeces, setting them aside for fleece competitions at the spring shows.

As the mountain of wool bags shrinks, a stack of boxes full of skirted fleeces takes its place. I tackle the remaining pile, a bit more each day, until the entire lot is boxed and bound for the spinnery.

SPRINGDELLE JACKET

Designed by Lisa Lloyd

Holding the freshly sheared fleece of one of my sheep is an invitation to dream of the yarn to come. The yarn used for this project, my Upland Wool and Mohair, is spun in a rustic, woolen style. It reflects the character of the hills and the diversity of our flock (a blend of Border Leicester and fine-wool crosses, with a dash of mohair).

This elegant jacket is stylish enough for a night out in the village while durable enough for lugging groceries. The mohair infusion of the yarn enhances the cables. The Springdelle stitch reflects the tidy newly sown fields of spring.

Finished Measurements

- Chest: 38 (40, 42, 44, 46)" (96 [101.5, 107, 112, 117] cm)
- Length from shoulder: 19 (20, 20½, 21)" (48 [51, 52, 53] cm)

Yarn

- Foxfire Fiber & Designs Upland Wool and Mohair (70% wool, 30% mohair; 140 yd [128 m]): 10 (10, 11, 11, 12) skeins. Shown in Aster.

Needles

- One 16–24" long circular needle size US 6 (4 mm)
- One set size US 4 (3 mm) straight needles
- One 29" (74 cm) or longer circular needle size US 4 (3 mm)

Notions

- Cable needle
- Stitch holders
- Stitch markers
- Tapestry needle
- 7 buttons, ⅝" (12 cm) diameter

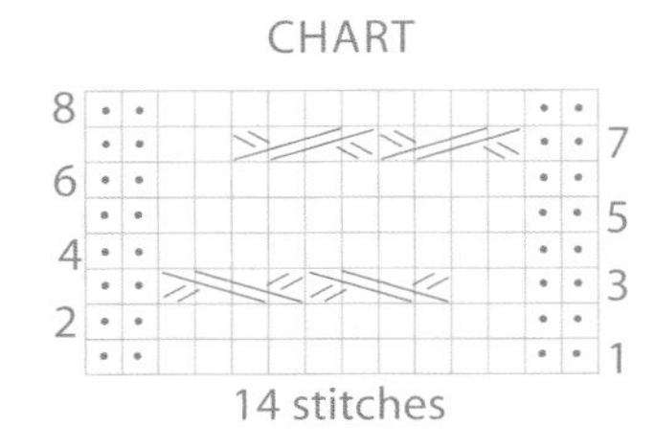

KEY

= K on RS, p on WS

= P on RS, k on WS

= Slip 2 sts onto cn and hold in back, k2, k2 from cn

= Slip 2 sts onto cn and hold in front, k2, k2 from cn

Gauge

20 sts and 28 rows = 4" (10 cm) in St st. Adjust needle size as necessary to obtain correct gauge.

Springdelle Stitch

Row 1: K1, *yf, sl 1, yb, k1, rep from *.
Rows 2 and 4: Purl
Row 3: K2, *yf, sl 1, yb, k1, rep from *, end k1.

Note: First and last stitches of each row are worked as a stockinette stitch selvage throughout. Lower edging is worked after the cardigan has been assembled.

BACK

With larger needles, CO 95 (101, 105, 109, 115) sts.

SET UP BORDER PATTERN

Row 1 (RS): K1, work Springdelle Stitch over next 93 (99, 103, 107, 113) sts, k1.
Row 2: P1, cont in patt until 1 st rem, p1.
Repeat rows 1–2 until border measures 3¾" [9 cm] from cast-on edge, ending with row 2.
Row 3: Knit.
Row 4: P1, knit until 1st rem, p1.
Repeat rows 3–4 once. Work in St st for 8 rows, ending with a WS row.

SHAPE WAIST

Decrease Row (RS): K1, ssk, knit until 3 sts rem, k2tog, k1. Work Decrease Row every 10th row 2 (1, 1, 0, 0) more times, then every 12th row 0 (1, 1, 2, 2) times—89 (95, 99, 103, 109) sts. Work 9 (11, 11, 11, 11) rows even.
Increase Row (RS): K1, M1R, knit

until 1 st rem, M1L, k1. Work Increase Row every 12th row 0 (0, 0, 1, 1) more times, then every 10th row 2 (2, 2, 1, 1) times—95 (101, 105, 109, 115) sts. Work until Back measures 13 (13½, 13½, 14, 14)" (33 [34, 34, 35.5, 35.5] cm) from the beg, ending with a WS row.

SHAPE ARMHOLE

BO 7 (8, 8, 9, 9) sts at the beg of the next two rows.

Decrease Row (RS): K1, ssk, work in patt to the last 3 sts, k2tog, k1. Work Decrease Row every other row a total of 6 (7, 8, 8, 9) times—69 (71, 73, 75, 79) sts. Work even until the armhole measures 7½ (7½, 8, 8, 8½)" (18 [19, 20, 20, 21.5)] cm), ending with a WS row.

SHAPE SHOULDERS AND BACK NECK

Row 1 (RS): Work 18 (19, 19, 20, 20) sts, attach a second ball of yarn and BO center 33 (33, 35, 35, 39) sts for back neck, work until 6 (7, 7, 7, 7) sts rem, w&t.

Shape shoulders as follows using a separate ball of yarn for each:

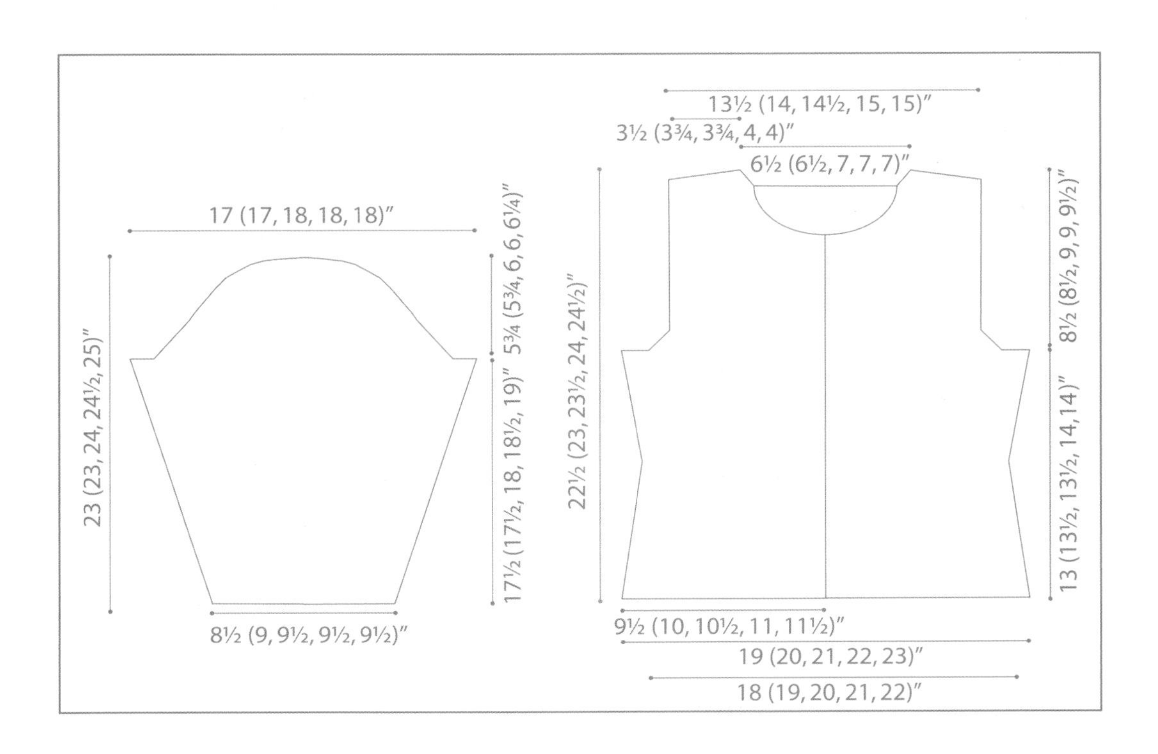

Row 2: Left Shoulder: work in patt; Right Shoulder: work until 6 (7, 7, 7, 7) sts rem, w&t.

Row 3: Right Shoulder: work until 3 sts rem, k2tog, k1; Left Shoulder: k1, ssk, work until 12 (13, 13, 14, 14) sts rem, w&t.

Row 4: Left Shoulder: work in patt; Right Shoulder: work until 12 (13, 13, 14, 14) sts rem, w&t.

Rows 5–6: Work across all stitches, picking up all wraps.

Place each set of shoulder stitches on holders—17 (18, 18, 19, 19) sts rem on each shoulder.

LEFT FRONT

With larger needles, CO 50 (53, 55, 58, 60) sts.

SET UP BORDER PATTERN

Row 1 (RS): K1, work Springdelle Stitch over 34 (37, 39, 42, 44) sts, pm, work chart over next 14 sts, k1.

Row 2: P1, cont in patts as established until 1 st rem, p1. Rep rows 1–2 until border measures approximately 3¾" (9.5 cm) from cast-on edge, ending with row 2.

Row 3: K1, knit to m, sl m, cont chart as established, end k1.

Row 4: P1, work chart as established, sl m, knit until 1 st rem, p1. Repeat rows 3–4 once.

SET UP MAIN PATTERN

Row 5: K1, knit to m, sl m, work chart as established, k1.

Row 6: P1, work chart, sl m, p to end. Rep rows 5–6 three times.

SHAPE WAIST

Decrease Row (RS): K1, ssk, work as established across the row. Work Decrease Row every 10th row 2 (1, 1, 0, 0) more times, then every 12th row 0 (1, 1, 2, 2) times—47 (50, 52, 55, 57) sts, then work 9 (11, 11, 11, 11) rows more.

Increase Row (RS): K1, M1R, work as established across row. Work Increase Row every 12th row 0 (0, 0, 1, 1) more time, then every 10th row 2 (2, 2, 1, 1) times—50 (53, 55, 58, 60) sts.

Cont until Left Front measures 13 (13½, 13½, 14, 14)" (33 [34, 34, 35.5, 35.5] cm) from the beg, ending with a WS row.

SHAPE ARMHOLE

BO 7 (8, 8, 9, 9) sts at the beg of the row, work as established to the end—43 (45, 47, 49, 51) sts. Work next row.

Decrease Row (RS): K1, ssk, work as established to end.

Work Decrease Row every RS row 6

(7, 8, 8, 9) times in total—37 (38, 39, 41, 42) sts. Work even until the armhole measures 5" (12.5 cm), ending with a RS row.

SHAPE FRONT NECK

BO 10 (10, 10, 10, 11) sts, work to end.
Decrease Row (RS): Work across to the last 3 sts, k2tog, k1. Work Decrease Row every RS row 9 (9, 10, 11, 11) times more—17 (18, 18, 19, 19) sts rem for shoulder. *At the same time*, when armhole measures 7½ (7½, 8, 8, 8½)" (18 [19, 20, 20, 21.5] cm) and with a RS row facing for the next row, shape shoulders.

SHAPE LEFT SHOULDER

Shape shoulder at outside edge as follows:
Rows 1, 3, and 5 (RS): Knit.
Row 2: Work to the last 6 (7, 7, 7, 7) sts, w&t.
Row 4: Work to the last 12 (13, 13, 14, 14) sts, w&t.
Row 6: Work across all sts, picking up wraps.

Place shoulder sts on holder—17 (18, 18, 19, 19) sts.

RIGHT FRONT

With larger needles, CO 50 (53, 55, 58, 60) sts.

SET UP BORDER PATTERN

Row 1 (RS): K1, work chart over 14 sts, pm, work Springdelle Stitch over 34 (37, 39, 42, 44) sts, k1.
Row 2: P1, cont in patt as established until 1 st rem, p1. Rep rows 1–2 until border measures approximately 3¾" (9.5 cm) from cast-on edge, ending with row 2.
Row 3: K1, work to m, sl m, k to end.
Row 4: P1, k to m, sl m, work chart as established, p1. Rep rows 3–4 once.

SET UP MAIN PATTERN

Row 5: K1,work chart as established, sl m, k to end.
Row 6: P to m, sl m, work chart as established, p1. Rep rows 5–6 three more times.

SHAPE WAIST

Decrease Row (RS): Work as established across the row until 3 sts rem, k2tog, k1. Cont working as established and at the same time work Decrease Row every 10th row 2 (1, 1, 0, 0) more times, then every 12th row 0 (1, 1, 2, 2) times—47 (50, 52, 55, 57) sts, then work 9 (11, 11, 11, 11) more rows.
Increase Row (RS): Work as established across row until 3 sts rem, M1L, k1. Cont working Increase Row every 12th row 0 (0, 0, 1, 1) more time, then every 10th row 2 (2, 2, 1, 1) times—50 (53, 55, 58, 60) sts.

Cont until Right Front measures 13 (13½, 13½, 14, 14)" (33 [34, 34, 35.5, 35.5] cm) from the beg, ending with a RS row.

SHAPE ARMHOLE

Next row (WS): BO 7 (8, 8, 9, 9) sts, work to end.
Decrease Row (RS): Work as established until 3 sts rem, k2tog, k1. Work Decrease Row every other row 6 (7, 8, 8, 9) times in total—37 (38, 39, 41, 42) sts. Work even until the armhole measures 5" (12.5 cm), ending with a WS row.

SHAPE FRONT NECK

Next row (RS): BO 10 (10, 10, 10, 11) sts, work to end. Work next row even.
Decrease Row (RS): K1, ssk, work until 1 st rem, k1. Work Decrease Row every other row 9 (9, 10, 11, 11) times in total—18 (19, 19, 20, 20) sts rem. *At the same time*, when armhole measures 7½

(7½, 8, 8, 8½)" (18 [19, 20, 20, 21.5)] cm) and with a RS row facing for the next row, shape shoulders.

SHAPE RIGHT SHOULDER

Row 1: Work until 6 (7, 7, 7, 7) sts rem, w&t.
Rows 2, 4, and 6: Purl.
Row 3: Work until 12 (13, 13, 14, 14) sts rem, w&t.
Row 5: Work across all sts, picking up wraps.
Place shoulder sts on holder—17 (18, 18, 19, 19) sts.

SLEEVES (MAKES 2)

LOWER EDGING

With smaller needles, CO 43 (47, 49, 49, 51) sts. Knit all sts for 4 rows. Change to larger needles.

SET UP BORDER PATTERN

Row 1: K1, work Springdelle Stitch over 41 (45, 47, 47, 49) sts, k1.
Row 2: P1, cont to work chart until 1 st rem, p1. Rep rows 1–2 until border measures approximately 4" (10 cm) from CO edge, ending with row 2.
Row 3: Knit.
Row 4: P1, k until 1 st rem, p1. Rep row 3.
Next row: Purl all sts and at the same time, inc 1 st—44 (48, 50, 50, 52) sts.

SHAPE SLEEVE

Begin working St st over all sts and begin shaping sleeve as follows:
Increase Row (WS): P1, M1R, purl until 1 st rem, M1L, p1. Rep Increase Row every 4th row a total of 22 (14, 15, 15, 19) times, then every 6th row 0 (6, 6, 6, 4) times—88 (88, 92, 92, 98) sts. Work even until Sleeve measures 17½ (17½, 18, 18½, 19)" (44.5 [44.5, 46, 47, 48] cm) from the beginning, ending with a WS row.

SHAPE SLEEVE CAP

BO 7 (8, 8, 9, 9) sts at the beg of the next two rows.
Row 1 (RS): K1, ssk, knit until 3 sts rem, k2tog, k1.
Row 2: Purl. Rep rows 1–2 a total of 6 (7, 8, 8, 9) times.
Row 3: Rep row 1.
Row 4: P1, p2tog, purl until 3 sts rem, p2tog tbl, p1. Repeat rows 3–4 a total of 2 (1, 1, 0, 0) times.

Rep rows 1–2 a total of 11 (11, 11, 12, 12) times.

BO 3 (3, 3, 3, 4) sts at the beg of the next 4 rows.

BO rem sts.

FINISHING

Join shoulders with a three-needle bind-off. Sew in sleeves and sew side seams.

LOWER EDGING

With smaller, longest circular needle and RS facing, pick up and knit 171 (180, 189, 198, 207) sts evenly around lower edge of cardigan. Knit 3 rows, then BO loosely in purl.

COLLAR

Note: Collar is folded over and stitched down to inside neck seam. A front edging is worked along the entire center fronts, stitching the ends of the collar closed.

With larger circular needle and RS facing, pick up and knit 94 (98, 102, 104, 104) sts evenly around neck edge.

Next row: Purl.

Work in Springdelle Stitch patt until until collar measures 1½" (3 cm).

Knit the next four rows to create a turning ridge.

Work in Springdelle Stitch patt until collar measures 4" (10 cm). BO all sts loosely. Fold Collar along turning ridge to inside and sew neatly to inside neck edge.

BUTTON BAND (LEFT FRONT)

With smaller circular needle and RS facing, pick up and knit 10 (10, 10, 10, 10) sts along edge of Collar, going through both layers, pm, pick up and knit 86 (88, 88, 90, 92) stitches evenly along the front edge. Knit 3 rows, then bind off all sts loosely in purl.

BUTTONHOLE BAND (RIGHT FRONT)

With smaller circular needle and RS facing, pick up and knit 86 (88, 88, 90, 92) sts evenly along the front edge to lower edge of Collar, pick up and knit 10 (10, 10, 10, 10) sts along edge of Collar, going through both layers. Knit one row.

Buttonhole Row (RS): Knit 6 (7, 7, 4, 5) sts, *BO 2 sts, k7 (7, 7, 8, 8) sts; repeat from * 5 times, BO 2 sts, k 18 (19, 19, 18, 19) sts, work to end.

Next row: Knit across row, casting on 2 sts over the bound-off sts.

BO loosely in purl.

Sew buttons opposite buttonholes. Weave in all loose ends. Hand wash and block gently to measurements.

LAMB WATCH

As March wanes, lamb anticipation heightens. I look at the ewes before leaving the farm for errands and do one late-night check. Although the calendar tells me the first lambs should arrive 145 days from the date the ram was first placed with the ewes last fall, my flock has taught me to stay on my toes and to expect the unexpected.

Moody and ravenous, expectant ewes, or "mothers," as we call them even before their babies arrive, are the size of zeppelins. At feedings they burn through hay rations in no time, shoving each other for every last blade, then looking for more. Ripening lambs turn their mothers into eating machines tapped out by lamb fetuses in their largest stage of growth. Like oversized carry-on luggage on a crowded flight, unborn lambs take more than their share of space in their mother's under-the-seat compartment, crowding the heck out of everything else. It's no wonder the girls are irritable, smashing each other like bumper cars in the barn. Shifting the balance of feed while maintaining the same nutritional plane helps ease them through the last four weeks of gestation, but if it's not done carefully, watch out. It can wreak metabolic havoc.

I begin subbing a ration of high-protein grain for a portion of the hay feeding. They love the grain, and though it's high in protein, it's also less filling. They would still gorge themselves on hay, given the chance.

With the sheep this far along, it's time to get my birthing kit in order. When you're delivering lambs at all hours, there's no time to go scavenging about the barn for the iodine bottle. A plastic caddy holds the items I want at hand when assisting a ewe: iodine, a bulb syringe for clearing airways, scissors, a bottle of gel lubricant. I search the cupboard in the former milk room, which holds everything a shepherd could possibly need in the course of a year, for my birthing supplies. As I take stock, I see that the top shelf holds a jumbled wad of nylon halters, half-empty tubes of paste wormers, and livestock marking crayons in every hue. An empty case for a digital thermometer (but no thermometer to be found) sits beside a mismatched pair of work gloves. On the shelf beneath sits the sling and scale for weighing newborns, a grubby bottle of iodine, a box of ear tags, one chipped ceramic mug, an empty yogurt cup, and a package of zip ties. I sort and tidy these shelves when the rare urge to organize strikes me, but order is fleeting on my farm. So I stop now and take stock, making note to order a thermometer and a new prolapse retainer.

The studio, the old farmhouse cottage just steps from the door of the birthing barn, becomes my base camp for the duration of lambing. When the season is in full swing and lambs are popping out at all hours, proximity to the barn saves time, and time is work. Or sleep.

In the tiny dining room, my shelves are lined with books on sheep management, the chapters on lambing well flagged. Sheep midwifery is something shepherds learn from experience; from talking to other sheep owners; or from books, the Internet, and magazine articles. There are no classes in how to be a sheep doula.

When I was first starting out, Paula Simmons's *Raising Sheep the Modern Way* was my bible. I remember closely studying the birthing chapters, as if memorizing every possible lamb malpresentation could ward off ever having to actually contend with one. Armed with optimism, not knowing how much I really didn't know, I envisioned lambing as a pastoral vignette. There were no breech lambs or uterine prolapses in my mind's eye.

While I've collected many resources for lamb survival, there is no true survival guide for shepherds. Over the years, I've developed my own checklist. I stock my studio cupboard with coffee, green tea, granola, and chocolate bars. My coveralls are on the coat hook in the mudroom, right above my barn boots. I gather long johns, wool socks, hats, an extra pair of gloves, and fresh batteries for the flashlight. An empty thermos waits beside the coffeepot.

No matter how prepared I am, there are times when a shepherd's best resource is another shepherd. Just short of that, the baby monitor on my nightstand and the lamb cams feeding into the TV next to the monitor become my most valuable tools for lamb watch. It takes a few nights to recalibrate my ear to the barn's soundtrack. The night barn is surprisingly full of sounds. The general ambient sound is the static hiss of the baby monitor punctuated by loud belches, followed by the rhythmic *crunch, crunch, crunch* of cud chewing. Ewes restlessly pacing through the straw from one end of the barn to the other make a sound remarkably like that of a hoof raking at the straw midcontraction. Gastric outbursts sound deceptively similar to the groans that accompany pushing. If my ear didn't learn to tell what was what, I'd never sleep at night. There's the electric moment when I'm awakened by a sound that's distinctly different. The lambs are coming.

The lamb cam was the ingenious idea of my friend Chris. Seeing how tapped out I was midway through lambing season one year, Chris offered to barn-sit one evening so Mike and I could grab a meal in the village. While minding the barn that night, he observed that my number of trips to the barn could be trimmed significantly if I were able to *see,* as well as hear, what the sheep were up to. Although it was a thought that had crossed my mind more than once (while numbly stumbling to the barn in my pj's at two A.M.), following through with investigating equipment was not a task that ever made it close to the top of the long list of priorities during lamb time (and was honestly forgotten about once lambing ended each year).

So I was truly surprised when Chris arrived several days later with a small wireless camera-transmitter-receiver set he had found while poking around on eBay. I couldn't wait to see if a clear signal could make it through the cinder block walls of the birthing barn to the little TV on my nightstand.

Since the camera was designed for monitoring a child's nursery, not a barn of expectant ewes, there was no easy way to mount it. We balanced it on an upside-down bucket to get a view of the aisle. The camera required leaving at least one set of lights on to produce a grainy but reasonable-quality image of the scene in the barn. The field of view was narrow, but an adjustment to the configuration of the pen kept the girls in the camera's line of sight. When awakened by an odd noise every now and then via the baby monitor the next night, I was elated and ever so grateful that rather than levitating out of bed and rushing to the barn at every sudden sound, I could tell that all was well in the barn without taking my head from the pillow and return to sleep.

The "Chris Cam" lasted the remainder of that season. If a fifty-dollar nursery camera could project a grainy black-and-white image from barn to house, it was clearly worth my time in the off season to further research the options for live video from the barn. In perusing catalogs from sheep supply outfitters, I came across an ad for the "lamb cam"—a wireless camera built for the outdoor environment. Although pricey, the camera was designed to work in low-level lighting. Before the start of lambing in 2007, I decided to upgrade, and by the time the first lambs hit the ground, my new lamb cam was mounted to the barn wall, an antenna beaming full-color, live lambing footage to a larger television purchased especially for nocturnal barn surveillance.

Since then, we've added a second camera. Lamb cam one catches the area just inside the south door and a good shot to the northwest corner of the sheepfold. Lamb cam two, mounted on a pillar in the northeast corner of the barn, pans the center of the pen all the way to the southwest corner. If the girls are aware that we are ogling them around the clock, they don't seem to mind. Early in the season, before the novelty wears off, Holly and I eat pizza and watch the mothers on television. There's no cable television at the farm, so we take our excitement where we can get it.

There's almost nowhere for the girls to hide, unless they stand in the corner directly beneath camera one. I had one mother birth in that spot. Though it was the only spot I couldn't see, I could see every other sheep in the barn staring at whatever was happening in the blind spot beneath the camera. I now rely on the other sheep to clue me in to action in that corner of the barn.

I can't imagine going through lambing season without my lamb cams and the baby monitor in the bedroom. It's an example of the interface of technology (as primitive as my system may be) and an ancient practice of livestock husbandry. I'm hoping in my lifetime to enjoy high-speed Internet access at the farm, so Mike and I can watch a live-feed barn camera from the house, without having to camp at the fiber studio during lambing season. But until then, the lamb cams will suffice.

THE FULL LAMB MOON

SHEPHERD'S LORE says the gravitational pull of a waxing full moon draws lambs. Although this doesn't always bear out, all shepherds have their stories. Here's mine.

The evening of March 11, 2006, was cold and clear, with a waxing gibbous moon, reminding me that the full moon was still a few days away. I did a last barn check at ten P.M. The ewes were quietly chewing their cud, comfortably ensconced in fresh straw. Several pens held new lambs and mothers. I wished all of them good night and went to bed expecting a quiet night.

My sleep was interrupted by the quick, sharp rustling of a hoof scraping straw. It had been little more than an hour since I last checked the barn. I waited, listened, and there it was again. A quiet barn, except for one intermittent sound.

Dressing quickly—insulated coveralls, wool sweater and hat, down vest—I grabbed my flashlight and headed across the lawn to the barn. A wash of moonlight made a flashlight unneeded. With my birthing kit in hand, I felt for the switch for the overhead lights just inside the barn door.

Fern, a Border Leicester ewe, was nestled in the straw, straining intently, with her head turned back over one shoulder. In the time it took me to walk to the barn, she'd transitioned from the restless discomfort of early labor to the serious business of pushing out a lamb. Grabbing clean towels, I parked myself in the straw nearby and let her work. Between contractions, she blatted, but she quieted as each contraction peaked, holding her breath, tilting her head back, getting ready to push and strain. Sacs filled with clear fluid resembling jellyfish emerged with each contraction, bubbles in the membranes that encase her lambs. Within twenty minutes the tips of a tiny set of hooves, the soles shaped like raindrops, appeared, followed by the tip of a small white nose. Fern suddenly lurched to her feet. Letting gravity do some of the work for her now, she squatted and pushed mightily to deliver a gangly pair of front legs, followed by a narrow head and then shoulders. In two more contractions the lamb, slick, shiny, looking like it was encased in plastic shrink-wrap, slid to the straw, headfirst.

Thirty minutes later a healthy set of twins shivered in the straw by Fern's side. Carefully lifting both lambs, I coaxed Fern to follow me into a five-by-five-foot pen called a "lambing jug." Focused on her lambs, she entered the pen without hesitation. And while she tended to one lamb, licking her all over with lightning speed, I tended to the twin, dipping its trailing umbilical cord in a cup of strong iodine. Sticky, sodden, and dazed, the lambs mewled softly, sounding like a cross between a kitten and a human infant. Fern snickered a deep guttural response while lapping the gelatinous veil from their nubby fleece.

I'm always amazed by how much newborns learn within a very brief span of time. The first few minutes must be sheer sensory overload. The differences in light, temperature, sound, must be overwhelming, but they are programmed to get on their feet quickly. The first few minutes are tentative experiments in physics and gravity. The back end gets hoisted first, so the first few steps are generally from a kneeling position and often in reverse.

Fern's lambs, a pair of ewes, were fast on their feet. In a series of steps in the art of differentiation, they quickly learned forward from reverse, how to tell mother from a water bucket, and front from hind. Then came the mechanics of vocalizing, rooting, latching, suckling.

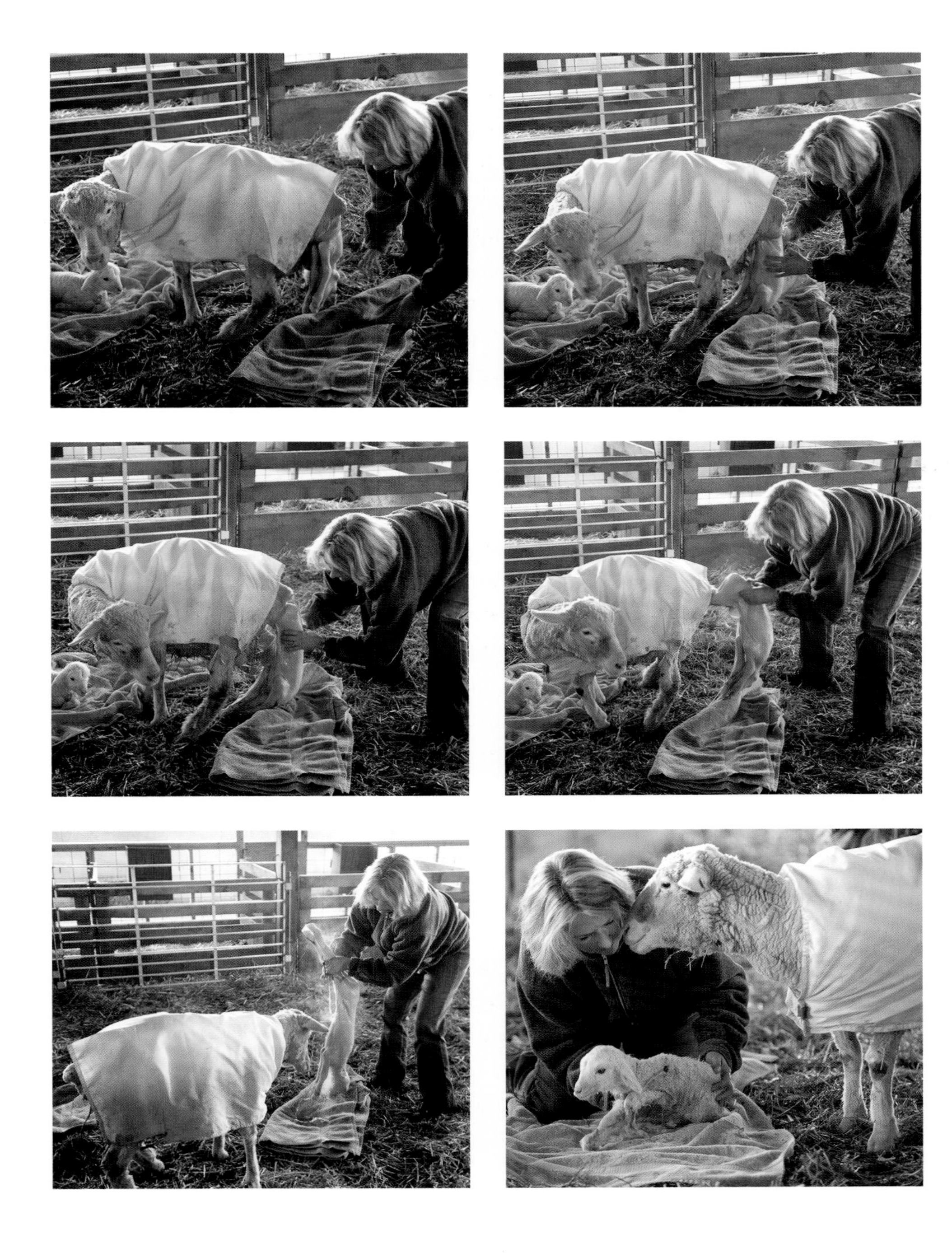

Instinctively the pair explored Fern's undercarriage, homing in on the udder. Smart lambs. Fern snickered while flicking her tongue over their tails and haunches, guiding them in the right direction. Good mother.

Once latched, the suckling reflex kicked in. Potent and laced with antibodies, colostrum is the thick syrup first produced by the udder. It's the equivalent of rocket fuel for lambs. If a lamb is slow to nurse, I'll strip a few ounces from the teat and use a small plastic syringe to squeeze a few drops into the lamb's mouth, to whet its appetite. It encourages them to get up and discover there's more where that came from.

Once certain Fern's lambs had learned where mother keeps the milk, I gathered the sodden towels and my kit. Preparing to exit the pen, I spied another Leicester ewe, Amethyst, in the corner with her head down, ears tilted back. She pirouetted 360 degrees and again stood head down facing the wall. As I fetched a bucket of fresh water for Fern and kept an eye on Amethyst, I noticed another ewe, Jewel, drawn to the wet spot in the straw where Fern had dropped her lambs an hour earlier. Rolling up my sleeves, I realized it would be awhile before I could return to my bed.

An hour later I was on my knees again in a lambing jug. I supported Jewel's ten-minute-old white ewe lamb while the lamb's much larger sibling, a black ram, wasted no time in chugging down more than his share of colostrum. The white eweling was having trouble working her legs but no problems at all exercising her vocal chords. She was tiny. And hungry. And *loud*. From her mother's udder, I milked a little colostrum into a yogurt cup and fed her with a small syringe. Like a shot of espresso, the syrup quickly energized her. She found her footing and shouldered her way to Jewel's udder.

My attention was next diverted by the new arrivals in pen number two. Amethyst's twins, who arrived just ahead of Jewel's twins at midnight, were looking for seconds, but Amethyst was distracted as she writhed while passing the afterbirth. With a trowel I fished the stringy blob from the pen, chucking it into the manure bucket.

Belle, a demure Cormo ewe, poked her nose through the wooden slats of the adjacent pen, checking out the newcomers with an unusual amount of interest.

And I then spotted Amy, a dark Leicester ewe who was standing alone in "the corner."

By two A.M. three ewes were restlessly milling around the common pen like planes circling in a holding pattern over a busy airport. Amy and her five-minute-old triplets were in the pen adjacent to Belle, with her pair of ram lambs. The barn now rang with the shrill cries of newborns and the guttural reassurances of attentive mothers. Not bothering with gates, I clambered over the wooden panels, moving from pen to pen. With birthings closely spaced, I was frustrated that I was not able to spend as much time as I'd like with each set of new arrivals. The barn was cold so I kept moving. I saw that each lamb was rubbed dry and had reached the udder before heading to the next pen. I found myself wishing for a thermos of hot tea. And some help. And then it occurred to me to call for help.

Within the hour Mike was in the barn, hauling fresh water to the lambing jugs while I tended to Sugar and her single lamb. Jewel's white eweling looked cold. I was not, so I wrapped her in my down vest and tucked her up against her momma.

I looked up to see the eastern sky brightening just as Asa, my favorite Leicester ewe, parked herself beside the row of pens, head cocked back over her shoulder, softly snickering to her unborn lambs. Hearing Kodiak's ewe lamb call from a pen and in a confused hormone-incited panic of misplaced maternal drive, Asa looked for a way to scramble into the pen to claim Kodi's lambs. Kodiak became irate, stomping her hoof in warning and punching at the pan-

els with her head. Draping a dhurrie rug over the panel to block Asa's view of Kodiak's lambs, I told poor Asa her lambs would soon be here too.

Thirty minutes later Asa's lambs were coming on like an express train. Pointed tips of hoofs emerged from her distended vulva with each push but receded as each contraction subsided. She strained with everything she had, legs outstretched, head lifted, neck arched, lips curled. The lamb was coming head and front feet all at once, creating a logjam. Grabbing the toes of one foot, I rotated the lamb slightly, easing one shoulder forward. The change in geometry freed the logjam, and Asa eased her lamb onto the straw. I then realized I had nowhere to put Asa and her twins. We were out of lambing pens. Mike and I improvised, lashing panels and odd gates together with baling twine. They jutted out into the aisle at funky angles.

There's little point in heading to bed when you've been at it all night in the barn and suddenly find it's nearly time for morning chores. Luckily, second string arrived. My barn assistant, Tish, offered to finish up so Mike and I could catch some rest. Before I could drink a cup of coffee back at the house, she frantically called to announce that Snowdrop was now in labor. I dumped my mug of coffee into a thermos and return to the barn.

Late morning, I was beyond exhausted. My overalls smelled of saline and were plastered with bits of straw, colostrum, and birthing goo. My hands were chapped and raw and stained with blood and iodine, my once-orderly barn was wall-to-wall lambs. The east pens were full, as were the impromptu pens set up everywhere else, seemingly arranged by a sleep-deprived maniac. Sodden towels were strewn over panels, and the various items from my birthing kit were scattered about.

Ahead of the full moon, half the flock had lambed in just over twelve chaotic hours. Now the lamb storm had passed. In a zombie-like fog, I looked into each pen, patting lamb bellies to see who'd nursed, accounting for all placentas. Unsettled by disarray, I pulled out my notebook, jotting down who'd had what, making notes of color and gender. The big picture was messy on the face of it. But as I looked at each pen, I saw a perfectly contented ewe tending lambs, in singles or pairs. They minded not that the barn was trashed; they had everything they needed.

MAX0104

JAMMIES FOR LAMBIES

Designed by Holly Sonntag

All lambs are born wearing sweaters, a nubby fleece that covers their pink skin. But for a set of twins born early one cold March morning, their natural sweaters weren't quite enough. That is how this project came to be.

I was in the pen with Chablis, a seven-year-old ewe, when she delivered two very frail ewe lambs just after dawn. Even with doors and windows closed, the barn was raw and drafty. The ewelings were undersized and shivery. Even after Chablis had licked them dry, they stood quaking, with their backs hunched up. I tried a brisk towel rubdown.

When newborn lambs are cold, they put whatever energy they have into trying to warm themselves. Cold lambs are not strong enough to nurse. I slipped my pinky (a handy oral thermometer) into the mouth of one lamb. The situation was iffy.

I took off the oversized gray wool sweater I was wearing over my coveralls and tucked both lambs together, swaddling them in wool. Their little frames needed insulation—and I remember thinking they could really use their own sweaters. That's when it occurred to me that I might convert the sleeves of my barn sweater into a pair of jammies. I unwrapped the lambs and lopped the sleeves from my sweater just above the elbow. Then I sliced a pair of holes just wide enough for their spindly front legs; it was my very first steek, although at the time I had no idea what a steek was. The lambs were now wearing their own cozy turtleneck sweaters. I patiently massaged each lamb out of its torpor, using a small plastic syringe to administer droplets of a tincture of molasses laced with vitamins and colostrum stripped from their mother's teats. Without having to work so hard at staying warm, the lambs eventually found their footing and their appetites, at last.

Putting sweaters on lambs shouldn't be necessary. Lambs are born with wool, and usually that, along with a good tongue washing from their momma, is all they need. Messing around with sweaters could actually be *a very bad idea*, especially with a high-strung, first-time mother who might be put off by a change in her lamb's appearance or scent. You don't want to impair the bonding process, but if a lamb is hypothermic, a little insulation goes a long way. And as I always say, a bottle lamb is better than a dead lamb.

When my original gray lamb sweaters became unraveled from use and washing, my blog readers had a sweater drive one year—sending care packages of sweater sleeves for my new flock's newest arrivals.

Eventually, using the former sleeves of my barn sweater as a model, Holly and I brainstormed a design for the perfect lamb sweater. Holly engineered the leg openings and had the cool idea to use scraps of yarn for color-tipped edging—to make it easier to tell newborn lambs apart.

Her Jammies for Lambies sweater is perfectly sized for a newborn lamb and can easily be modified to fit a small dog.

Finished Dimensions

- Length: 12" (30.5 cm)
- Girth at widest circumference: 15" (38 cm)
- Circumference at neck ribbing: 10" (25 cm)
- Circumference at waist ribbing: 14" (36 cm)
- Note: To adjust the pattern for larger (or smaller) lambs or other four-legged creatures in need of pj's, substitute bulkier yarn for larger animals or a sport-weight yarn for smaller critters.

Yarn

- Foxfire Fiber & Designs, Upland Wool and Mohair (70% wool, 30% mohair); 140 yd [128 m]: 1 skein main color (MC)
- Scrap yarn in a contrasting color: approximately 8 yd [7.5 m] (CC)

Needles

- One set double-pointed needles size US 7 (4.5 mm)

Notions

- 4 stitch markers, each a different color
- Tapestry needle

Gauge

- 18 sts = 4" (10 cm) in St st. Adjust needle size as necessary to obtain correct gauge.

JAMMIES

With CC, CO 60 sts.

Put 24 sts each on dpn 1 and dpn 2. Leave rem 12 sts on dpn 3.

Place marker and join in round, being careful not to twist.

Change to MC and work k2, p2 ribbing for 3" (8 cm).

Split for legs and work in stockingnet st back and forth. Using only sts on dpn 1 and dpn 2, work as follows:

Row 1: Knit.

Row 2: Purl.

Rep rows 1–2 for a total of 12 times.

Break yarn. Using MC knit only sts on dpn 3, work as follows:

Row 1: Knit.

Row 2: Purl.

Row 3: K1, M1, k10, M1, k1.

Row 4: Purl.

Rep rows 1–4 for a total of 5 more times—24 sts on dpn 3.

Next row: Knit.

Next row: Purl.

Join dpn 3 to dpn 1 and dpn 2 and cont working in rounds.

Knit for 3" (7.5 cm)—72 sts.

Work in p3, k3 ribbing for 2" (5 cm).

With CC, BO all sts and pull end through last loop. Use tapestry needle to weave in ends.

MAX0104

TROUBLESHOOTING LAMBS

ONCE WE'VE "LAMBED OUT," there's a tremendous feeling of relief. The drama of birthing, with its attendant worries, is behind us for the season. But with a nursery of newborn lambs comes a barn full of new responsibilities. In the space of the four weeks it usually takes for lambing, my work has quadrupled. Caring for a barn full of lactating ewes and their new lambs really ups the ante. I've learned the pitfalls, major and minor, and how best to avoid them.

At the start of each season, I swear I will not raise a bottle lamb. In shepherd parlance, bottle lambs are also called "bummer lambs." I've never searched the reason behind this name. Hand rearing lambs is superdemanding. Mixing lamb milk replacement, toting bottles to the barn every three to four hours around the clock, washing bottles: with a barnful of new charges, the last thing I want is to be wet nurse to a string of orphans—no matter how adorable. The extra work truly is a bummer. And yet, although I try my hardest to use the bottle as a last resort, we invariably have bottle lambs each year.

It's uncommon for a ewe to reject (or to be unable to feed) her own lamb, but it can happen. A first timer with a set of twins but only enough colostrum for one selected lamb to nurse will pointedly avoid her second, the smaller of the two. Ewes will also reject lambs for other reasons. Sometimes, if the lambs are of different colors, the mother discriminates by sight. Helena, a crossbred Cocoa daughter, confounded me two years in a row because she didn't like the "look" of one of her lambs. In 2009 she rejected her white eweling Blaze in favor of her black lamb Cinder. And then again in 2010, she pulled the same stunt, this time rejecting her black ewe Mistral in preference for her ram Matisse. After those experiences, Helena took an early retirement from her career in motherhood.

If the lambs are the same color and nearly equal in size, the mother may dis-

criminate by scent. I smear Vicks on the nostrils of the mother and all over the butts of both her lambs if I suspect a ewe is questioning the legitimacy of her own lambs. Sometimes a ewe detects that something isn't quite right with one of her newborns. She may treat them alike at first but then slowly come to her own quiet conclusion that nature has gone wrong, and then she will pointedly avoid the reject.

Some bottle lambs are my fault, the result of poor decision making. I came to this realization in the spring of 2008. It was late season, when my reasoning had unraveled. A newborn lamb was off to a rough start. Although close in size to his sibling, he was weaker from the beginning, too shaky to support his own weight. His brother was robbing more than his share of what little colostrum was available. And though I was loath to do it, I stomach tubed the lamb, sliding a flexible plastic tube down his throat and into his stomach and adding a few ounces of milk. This trick usually jump-starts a weak lamb. But even after several stomach-tube feedings throughout the day, this little guy didn't rally. Naming him Issey—though Iffy might have been more appropriate, a friend suggested—I worried throughout the day. It seemed unlikely that he would survive another night. And so we started Issey on the bottle. As we had suspected, once we had started bottle-feeding Issey, his first-time mother wouldn't have him back. During his first few days, Issey was content to sleep wrapped up in a towel in a rubber grain tub on my kitchen floor. But nighttime was a dilemma—where to keep one orphan lamb?

A potential answer seemed to present itself the next day when Lavender, another first-time ewe, went into labor. I had heard that it is possible to graft a motherless lamb onto a ewe *if* the lamb is young enough and *if* you are able to slip the foster lamb into the ewe's presence as she sees her own lamb for the first time. The process is officially called slime grafting, since an important step in getting the ewe to accept a lamb that isn't hers is to rub as much birthing goo as possible on the orphan so that both lambs are slimey and, one hopes, smell alike.

As Lavender's lamb was born, Mike deftly passed Issey to me. I quickly bathed him in the warm, sticky saline-smelling amniotic fluid that came with the newborn lamb. When Lavender turned to see her new lamb, she saw instead *two* lambs. She was skeptical from the start. Either I hadn't slathered enough goop on

Issey or Lavender was just smart enough to sense that something wasn't kosher. In cleaning her lamb, she pretty carefully avoided making tongue contact with Issey, despite my attempt to position the two lambs in proximity.

Then she delivered a second lamb.

Lavender began cleaning all three lambs, and then it slowly seemed to register that something still wasn't right. Her focus went back to her firstborn lamb. No matter how hard I tried to get her to clean her second lamb, she passed right over him. "Born-again" Issey was by this time on his feet and had found her udder. Having a twenty-four-hour age advantage, he wasted no time in helping himself.

I placed the three lambs beneath the warming light, thinking the sight of them all together in her pen would warm her heart. . . .

She didn't buy it. Inspecting them all closely, she quickly identified her favorite—her first lamb—and once again ignored the other two. She then became more forceful in expressing her feelings about the two that (she thought) weren't hers, head bashing them each time they stood.

I then began to realize I had made a pretty big mistake. Not only did the slime graft fail, but I had also confused a perfectly able first-time ewe into rejecting one of her own lambs. I now had two orphans on my hands . . . which, given the late hour, meant two houseguests for the night. We placed Issey and "the other guy" (I was too tired to name the poor thing) in a box beside our bed. The lambs, who seemed comforted by the barn sounds coming over the baby monitor, woke hungry about every two hours, so I bottled them. Partway through the night, they became restless. I let them out to stretch and walk around so they could make themselves sleepy again, and I spent what was left of the night *not* sleeping on the floor, where I could keep tabs on two now-adventurous lambs bumping about the bedroom. I was afraid one would stick his nose in an electric outlet or pull the clock radio off the night table. It was a very long night. Each time a ewe's call came over the baby monitor, the lambs answered.

Thank goodness I didn't have to go to the barn again that night to deliver more lambs.

Beyond the bottle lambs, each day in the lambing nursery starts with triage. One day during lambing season, one little ram lamb from a set of twins was

unhappy, looking cold and hunched over, a sign of hunger. I scan the ewes at the feeder until I spy his mother mowing down breakfast. Then I find his sister. Scooping up the pair, I place them near their mother and watch. The little ewe immediately dives and punches away at the udder. Her brother punches the teat on the other side and tries to latch on. Mother's not having it. She lifts her leg and abruptly shakes him off. He gives it another shot, but this time mother kicks him off a little more emphatically. Time to troubleshoot. I scoop up the ram lamb again and run my finger inside his mouth along his lower gum. Lambs are born without teeth, but within days their milk teeth erupt. The little ram lamb has a mouthful of pointed daggers. No wonder his mother doesn't want any part of him. I pull out the emery board I keep in my vest pocket just for this reason and with the lamb on my lap perform a little dentistry. By gently filing the corner of each tooth, I remove the sharp edges, testing it with the tip of my pinky. I then find his mom, still at the feeder. Pinning her against the feed bunk with my hip so she can't take off, I bend to examine her teats. Sure enough, she has a few raw-looking gashes where this little guy has done a number on her. I massage her udder with some Bag Balm and give her a handful of grain with some chewable aspirin mixed in to ease her discomfort. I make a note of the lamb's and the mother's ear tag numbers on the barn log, the daily chronicle of concerns.

Another pen holds the sets of triplets. Some years there are no triplets, but this year's arrival of three sets within a week of each other has made it necessary to set up a triplet nursery so the mothers can have an extra ration of grain. The general guide is a pound of grain per lamb, once the lambs have really got the hang of nursing. By helping the mothers support their triplets, we hope to avoid bottle lambs if possible. But one triplet is not looking as solid as her larger, more-robust siblings, and I've had my eye on her for a day. It's time to see if she'll take a few ounces of lamb milk replacer from a bottle.

Mornings always include a tour of the lambing jugs, inspecting the newest arrivals, assessing needs, dealing with the crises du jour. Ewes mind my handling of their lambs less if they're distracted by breakfast. But some ewes are downright fractious. Holly and I developed a shorthand for noting the temperament of the mothers. *Stompy* became our code word for the superfierce mothers, who would stand between us and their lambs and stamp a front hoof at us when we entered their pen. Holly and I would leave each other notes on the whiteboard near the lambing jugs: "Caution—Pansy is *stompy*." A really cranky ewe was labeled "Extra Stompy!"

Standing over a pen that holds twins born to Tansy just before daybreak, I see that they are on their feet, mouthing a few stems of momma's breakfast. The sun is streaming in, flooding their pen in warmth, so I remove their sweaters. I fish around in the straw for the stringy afterbirth, tossing it into the refuse bucket.

By the time I've made my rounds and tended to a series of mini–lamb crises, an hour has passed. I've fed and watered the mothers, but the rest of the flock in two other barns needs food and water. And then the lambing jugs need to be mucked out, mineral feeders need filling, birthing towels need laundering. I've been up since 3:30 A.M. and have somehow misplaced my thermos of coffee in shuffling about my chores. I assess my own needs: breakfast, caffeine, or sleep.

By the time I finish chores and make it back to the studio, sleep wins out. Not bothering to change out of the jeans and tattered flannel shirt I wore to the barn, I crash until lunchtime.

Lambing season takes its toll. Clothing becomes untucked and mismatched. Two clean socks make a pair, regardless of color or pattern. Housekeeping and hygiene slide.

I have to remind myself to stop and appreciate the exuberance of a barn brimming with new lambs. Sitting on the concrete stoop at the south end of the birthing barn one sunny morning, soaking up the sun, I'm still enough to become part of the woodwork, and a raft of lambs comes to float in the warm pool of sun beside me. Behind me, in the barn, there's poop to scoop, wheelbarrows to dump, feeders to sweep, buckets to fill. But the flock is at ease, and so I, too, accept that for the moment, everything is as right as it can be. I share the pool of sun with my lambs, even though the napping bottle lamb in my lap has just piddled on my jeans.

LAMB-PEDES AND A LLAMA

AFTER THE FOUR-WEEK WINDOW of birthing time, I'm back to sleeping in my own bed at night. "Last call" for the bottle lambs is usually around eleven P.M., the best time to step into the barn and spy on the secret life of sheep. The mothers have crashed in the straw like beached seals. Some lambs are tucked up close to their mothers. But there are always a handful who just aren't ready for bed. They paw at their mothers to make them stand or scramble up onto their mother's back. The llama is an object of fascination, a magnet for incurably curious lambs. Gathered around Crackerjack, who is lying sphinxlike in the straw, they pick at his scruffy fleece, tugging gently with their mouths. Cracker remains tolerant, inscrutable. Until one of the little monkeys scrambles onto his back. Crackerjack draws the line. Swiveling his long neck over his shoulder and flattening his ears, he feints a spit at the little hooligan. This remonstration is all that's needed. Moving so as not to create a stir, I head to the pen holding the bottle lambs.

Tending to the lamb nursery consumes several hours each day. But I don't mind because I have a front row seat to their antics. Their behavior patterns

change as they go through developmental stages, just as children's do. But as is the case with almost all animals, the learning curve is faster.

The newest arrivals explore the world through their mouths, seeking all potential food sources within the barn. Just for fun I toss a plastic ball into the pen. After a moment's alarm, the lambs can't resist investigating. What is it? Where are the teats? They push the ball with their heads.

The boldest lambs put the moves on the mothers lined up at the feeders in the morning, coming in from behind, snitching milk from all the ewes. You can spot the snitchers—the lambs with yellow foreheads. Urine stains are a telltale sign of the sneak approach from behind. A hungry ewe is too intent on hay to care which lamb is nursing, especially if she can't easily see who it is, so the

little ones get away with it. But if a mother smells or spies a lamb that isn't her own helping itself, the lamb gets a sharp rebuke.

As lambs figure out how to use their long and gangly limbs, they discover the art of bouncing. They propel themselves upward, backward, and sideways until they finally get the hang of bouncing in a straight line. Then they bounce from one end of the barn to the other while the mothers eat. The bounces go higher and sideways with wild ninja kicks. The aerial kicks get wilder and more exuberant. Once they get the hang of it, it's a competition of daredevils.

As the boys fill out, they engage in ram play, which is really the earliest stage of establishing the pecking order. They begin sparring, butting heads with each other, ears back, tough expressions on their faces. But they'll butt anything in the pen: buckets, balls, feeders, the rug hanging over the panel. As their personalities emerge, I begin to see which ram lamb really has the right attitude for future breeding. The most dominant lambs mount each other and harass the poor ewelings. It's all play at this stage, but they're already preparing for the serious work of being sheep.

When we send the lamb flock outdoors for the first time, usually near the middle of April, as the grass is greening, they are wide-eyed with wonder at the vastness of space and sky. At first they stick tentatively to their mothers' sides. If they get separated, they *baah* hysterically until their mothers answer. Lambs and ewes recognize each other primarily by scent but also by voice. So the lambs' first few days in the yard are noisy as they learn to keep track of their mothers.

Eventually, the lambs begin stampeding. With barn doors wide open, they now race the entire length of the barn, launching themselves like rockets off the threshold of the barn door. In the yard they gallop like fillies in a loop before charging in unison back into the barn. They race closely in a pack and repeat the whirlwind race over and over again. The only way to avoid being trampled by the lampede is to join in. Even the mothers join the ewes. This tight flocking behavior is a survival strategy. Like so many prey animals, sheep learn to gather up, stay tight, and flee in unison. It's much harder for a coyote to pick off a single animal from a tight pack than a straggling sheep. I believe the flocking and the races are practice in evasive maneuvers for being out in the pasture later in life. But in the meantime, they lamb-pede for the joy of running and leaping in unison.

By the time the flock transitions from yard to pasture, the lambs have become nearly as ravenous as the ewes. Mimicking the behavior of their mothers, they quickly learn to relish the greenery and tear mouthfuls of turf. They scramble in tight formation behind the ewes. As the ewes fan out, the lambs get distracted and separated. Once they realize they're on their own, they'll sound off in panic until answered by their moms.

My llama, Crackerjack, wears many hats. He is a scruffy-looking, gray-freckled gelding with a goofy grin (caused by lower teeth that jut from his mouth). In the birthing barn he serves as doula to the mothers, who often position themselves in his proximity when they go into hard labor. He inspects every pen of new arrivals, usually sleeping beside the pen of newest lambs. In the nursery he is the lamb nanny, entertaining the lambs while the mothers nap. Now, in the pasture, he adopts the position of guardian and leader. Innately sensing that the lambs are the most vulnerable at this stage, he'll position himself near them, pausing while chewing a mouthful of grass to scope out the pasture, always watching for something out of the ordinary. I'll notice a handful of lambs, usually the oldest and most confident, gravitating to the llama in the pasture. Instinctively they recognize him as both leader and bodyguard. The instinct to follow him becomes more important in the weeks to come. Once the lambs leave their mothers' sides, they'll need someone to follow. When it comes to leading lambs or simply staying at their sides, Crackerjack shines. I count on him to keep the lambs out of trouble for when I *do* manage to get away from the farm.

SUMMER

In comparison with spring's unrelenting pace, summer works in syncopation with the weather, which always dictates what gets done when, and how quickly. The locus of action shifts from barn to field: sheep focus their energy on grazing; we shift our focus to the land on which they rely. Tending fence and field takes as much time and energy as tending flock.

Much of spring's activity happens in the shelter of the barn, no matter what the weather is outside. In summer it seems everything that needs doing requires a dry sunny day for getting done. Even with longer hours of daylight, between its late arrival and sometimes early departure, summer is our shortest season.

HAY-LELUJAH!

KEEPING SHEEP FED through a New England winter relies mightily on making hay while the sun shines in summer. First-cut hay comes off midsummer. The yield is almost always great, since the grass grows fastest in the early season. I've stood in grass nearly shoulder high in early July. The growth in the latter half of summer, the second cutting, makes the best winter sheep forage.

In western Massachusetts summer comes tentatively in May and June with a string of bright, clear days bracketed by periods of intense showers. Intervals of rain propel grassy stalks upward in both hay fields and pastures. While lambs merrily munch in the pastures on one side of the fence, the grass rockets in the hay field on the other side. In a perfect year, May rainy days yield to a dry week in mid-June for first cutting.

At a distance, hay making may look as simple as mowing an oversized lawn. But hay making is a science and a dance between farmer, turf, and sky. Grass-based farmers spend a heck of a lot of their time worrying about grass, especially during the growing season. Sometimes it seems as if it's all we think about.

Conversations with neighbors begin with comments about weather and then invariably segue into how the grass is growing. We rely on what's pushing up in the pastures to feed our flocks and herds in summer—and on whatever we can harvest from the fields to fill the feeders in winter months. In drought years, when the fields are toasty brown, we really sweat about the animals' finding enough forage. The hay yield is poor in those years, and we think carefully about the number of sheep we can support throughout the winter. Some summers bring monsoons—days and days of rain. The grass grows like crazy, but there's no opportunity to harvest. You cannot make hay in the rain. In fact, you need a solid week of good, clear, dry weather to make good hay. In the wet years all eyes are on the ten-day forecast, watching for that one window of opportunity.

In early summer, the clouds call the shots. On our farm there is almost no such thing as dry—there are degrees of wetness. Our hay fields get wet in spring and stay soggy largely because of a layer of clay not far beneath the soil's surface. It forms a liner. Walking across areas of hay mowings in June is like stepping onto a saturated sponge. Shaded by the ever-growing stalks, the turf at the base takes nearly a week of dry before you can even think of driving a tractor across the field. One pass, and the tractor tires are shiny. Next pass, and your wheels are wallowing in a grass rut.

Grass loves this time of year. But with half of our fields, the grass is virtually in standing water. We have to wait it out for first cutting. It's impossible to harvest hay on a wet field, so quite often our wettest fields have to wait until a good dry spell in late June before the first harvest. The longer we wait for a dry spell, the taller the grass. If the grass gets too leggy, its nutritional value drops. The trick is to harvest the grass before it bolts.

By the time first cutting gets under way, the shadows of the hay field at the edge of my garden towers taller than my own. The dogs won't set foot within the dense wall of leggy green stalks behind the house. Feathery heads of rye and slim and cylindrical wands of timothy dance in the wind. The optimum time for harvesting is while the grass is at its nutritional peak—the vegetative state before the seed heads let loose.

Timing is everything. And the more time that passes before first cutting, the less time remaining for second cutting. The sheep are indifferent to this important detail; grass is grass to them, whether at their feet in the field or dried and in the feed bunk. But I keep an eye on the ground and the sky and my ear tuned for the sound of tractor and sickle mower.

While a delayed first cutting may spell trouble for the sheep, it's beneficial for other farm residents. By mid-June many of the ground nesters have fledged. The does who bed their spotted fawns in our mowings for cover have learned to bolt at the sound of a tractor.

When it comes to hay making in general and first cutting in particular, we are bystanders in the process; it's our neighbors, the Davenport family, who have the know-how, the equipment, and the true feel of the dance. Norm gives me a heads-up when he sees a break of three or more dry days in the forecast. My job is to round up the sheep and keep them out of the way.

Many of our sheep pastures abut the hay mowings. The only thing separating

them are the rows of white electrified net fence strung earlier in the season to keep the sheep off the hay crop. On hay day, I round up the sheep after their morning graze, calling them into the paddocks close to the barns. It's where they prefer to spend their days anyway, loafing in the shade, and they are easily summoned.

Not so easy is pulling up the net fencing from the perimeter of the hay fields. The grasses have threaded their way in and out of the fence, a warp of tall green stalks woven through a weft of electrified poly twine. Taking up fence requires a solid tug on the white vertical struts supporting the net. Each tug pulls some grass along with it, and since every step requires another strong tug, this provides a workout that is both aerobic and muscle toning. It's not long before I'm trickling with sweat and breathing heavily, which alerts a cloud of blackflies (in an early first-cutting year) or a persistent deerfly escort (if first cutting comes late in June). Pollen shaken loose from the seed heads dusts my arms, and my jeans are soaked with morning dew above the boot line. Once the sheep have been rounded up and the fence rolled aside, first cutting gets under way.

Norm and his family make round bales with the first mowing. Since sheep prefer second cutting hay, Norm sells first cutting to a local horse farm. What doesn't go to the horse farm feeds Norm and Lisa's cows in winter. On some farms, round bales are all that are made. The five-hundred-pound round bales require less curing time and can be retrieved quickly by tractor, loaded on a wagon, and ferried to the hay barn, where they get stacked one atop another, four bales high. It's a huge time- and labor-saving method of putting up a lot of bales without the need for so many hands. A couple of people on tractors can clean up a round-bale mowing in a few days.

I've watched Norm round-bale for eleven years, and I still find the workings of the round baler mysterious and wondrous. Pulled behind the tractor and powered by a P.T.O. (power takeoff shaft), the baler resembles a giant steel clamshell. A bristle brush, not unlike the beater bar on my vacuum cleaner, forms the leading edge, where the baler scoops up the swaths of raked grass. The inner chamber houses a labyrinth of gears around which winds a rubber belt. The process involves many yards of baling twine per bale. The grass gets swirled and compacted into a giant coil and secured with twine. It all happens somehow inside the metal clamshell. When the bale is "done," the clamshell, which is hinged at the top, lifts and opens and deposits a newly minted round bale in the field.

I'm not a mechanical person. I have a clearer understanding of how to upload data to the Internet than I have of exactly how a round baler works. And I'm always grateful once Norm, Lisa, and Fred have pulled the last of the round bales from the field. What was days earlier a swaying sea of stalks and seed heads now looks brownish and bristly. Exposed to the air and sky, the earth dries out a bit, for the first time since the start of the season. As we restake the net fence and usher the flock back to summer pasture, their food for winter forage begins to emerge.

CUTTING THE APRON STRINGS

THE GREENING and growth of the summer pastures are timed well with the needs of the flock. Before we reach the summer solstice, I begin to notice that collectively the ewes' well of maternal patience is running low. By mid-June the lambs have quadrupled in birth weight and appetite. Using a socket wrench and a pair of vice grips, Mike has adjusted the spacing of the bars on the creep pen twice since the early spring to accommodate the increasing girth of our most roly-poly lambs, while keeping the spacing narrow enough to prevent the mothers from sneaking through. A "creep" is a special pen that allows lambs access via a creep panel, a gate with slots through which they must walk or creep. The slots are too small for the mothers to follow, but they try anyway. Inside the creep is a self-service grain buffet for lambies. The creep is a happening spot.

The pasture is plentiful and there's grain in the creep, but for the lambs nothing tastes better than their mother's milk. I watch a pair of forty-five-pound ram lambs duck beneath their mom and punch her bag to get the milk flowing. The impact sends her into a reverse wheelie, popping her hind feet clear off the ground. Their eager nosedive for the udder and her wearied expression tell me it's time to cut the apron strings.

Weaning the lambs is a headache. Timing is important, so it's not something I decide to do one day on a whim (unless I want an even bigger headache). When the lambs are between ten and twelve weeks old, they have developed rumens capable of processing grass and grain. I plot the calendar. Weaning should occur before or after (but not during) the first hay cutting. Not during a heat wave. And definitely during a week without travel plans.

Some shepherds cut the lambs off cold turkey, an abrupt approach that seems needlessly harsh for both lambs and ewes. Experience has taught me that any transition handled with time and patience—taken day by day, being mindful of the well-being of each group—pays off in the long run. Since any abrupt metabolic change can create a weakness in the wool staple, it behooves me as a wool grower to be especially mindful of this transition. So weaning happens in stages at our farm to minimize stress on ewes, lambs, and fiber.

It's difficult to "dry off" ewes who are on good pasture. As long as they're on lush grass and the lambs are nursing, they'll make milk. I'll start a week or so before weaning day by running the mothers on coarse pasture that's been pretty picked over. The lambs can come and go to their mothers as they please, but they can also access some better ground and their creep.

There's little complaint from the ewes, as this feels like a holiday. Although their pasture is somewhat meager, they're content to have it mostly to themselves while Crackerjack and the lambs go off on a picnic. When their udders get heavy and achy, they *baah* for the lambs. The young ones don't mind skipping back and forth between pastures. We let them come and go like that for a few days.

The ewes resent the next move: when we shut them in the barn and place a creep panel in the doorway. Again, the lambs can come and go, but the moms are now stationed inside the barn, where we transition their diet to coarse, dry hay. In winter months, it would be perfectly palatable, but in contrast to pasture, it feels like a diet of bread and water, and it does not play well. The ewes blat unhappily for their lambs or maybe for the green grass on which their lambs are grazing. The lambs ignore them. They've got grass and a llama at their side. Life is good. When the heat drives them back into the barn, they nurse. It goes this way for a few more days. Until the day comes when the lambs head back to the paddock to nurse and find the barn empty.

Pulling ewes from their lambs goes against the grain of their instincts. Weary and thin from ten weeks of lactation, they are still protective mothers. Nevertheless, while their darlings are out in the field, they can't resist the promise

of a grain pan. It's the only thing they want more than their own lambs, or at least they do at that very moment. After several days of a dry hay diet, they are easily led to the barn across the road.

I pay heavily for my deception. The next forty-eight hours are pure hell. When the lambs come bounding back to an empty barn, panic sets in. With eyes wide, heads pivoting this way and that, they call to their mothers. When they get no response, they scramble furiously back to the pasture to see if the moms have somehow been out there all along. "They must be there! How could we have missed them?"

Their blatting gets more insistent in volume and pitch. They begin pronging in anxiety back and forth between barn and field. Cortisol levels rise to the rafters, where the English sparrows chime in. Across the way, the mothers hear the fracas and begin to cry foul. Anxiety goes airborne. Frantic alarm now ensues from both sides of the road.

Crackerjack presides over bedlam in the lamb barn. Although it's difficult to read a llama's expression, I believe I see a look of resignation come over his face. Confronted with forty lambs pinned to the roadside gate blatting hysterically, he turns his head and looks at me as if to say, "And whose idea was this?"

He's been down this road before and knows it's going to be a long night. At evening chores, we make sure the gates are tight and the fences are hot. The lambs find solace in their evening grain. Although it's unnecessary, I pile a high-value dinner—the last of last year's best hay—into their feeders and close them into the barn for the night. The ewes, whose udders are well past nursing time, hoarsely bellow to the lambs.

Blatting lambs and bellowing ewes becomes the soundtrack for the entire neighborhood that evening. As dusk filters in and the day winds down, the yelling of lambs and ewes is amplified by the quiet. Everyone on the Patten knows the lambs are weaning. Full throated, they *baah* until hoarse. I sleep with the window open, listening for coyotes whose curiosity is undoubtedly piqued by the distress calls of young lambs.

One year the Wheelers, who raise Belted Galloways up the road for beef, weaned their calves the same weekend we weaned lambs. It was an Olympic bellowing competition, the Wheelers' Belties versus the Parrys' sheep. I'm sure the sound carried for miles.

The first night is always the worst. By morning the lambs are wrung out. Hunger pangs win out over absentee mothers, and they set about grazing with renewed intensity when we open the pasture gate.

The mothers, who will remain on a bare-bones diet for several days, still want their lambs back in the worst way. They loudly register their complaints. Over the course of gestation, lambing, and lactation, my role as gate mistress and controller of all resources has become firmly fixed in their minds. Their expressions say, "See here, there's been a terrible mistake!"

I see their bags full of milk and feel very sorry for them. The first twenty-four hours are hardest. Mike sets up fans in the barn and makes sure they have fresh, clean water—but not too much, as water augments milk production. It seems harsh, but it would be far worse for the mothers to develop mastitis. For the next week, we study udders daily for signs of angry redness or lumpy swelling. We allow them the coolness of the barn for refuge. By day three, there's peace again on both sides of Reynolds Road. On the west side of the road the lambs forage with Crackerjack, their nanny, at their sides. On the east side, udders go slack. In one week's time, the mothers, too, can return to pasture.

GRAZING DAYS

NEWLY WEANED LAMBS go through a transitional time before truly coalescing as a little flock within the flock. With the exception of the bottle lambs, they are wary of me, and without their mothers to guide them, they are rudderless.

Before we can move the lambs from pasture to pasture throughout the summer, they need to learn how to behave as a flock. Teaching them to flock up properly and showing them the lay of the land take patience and an understanding of how they both perceive and might respond to new situations. To do so I need to anticipate what may go wrong before something actually does go wrong and use the sheep's natural behavior to accomplish my goals with as little fuss as possible.

While it's fairly easy to flock up the lambs in the barn, handling them in the field is another story. Distractions are everywhere and the variables less predictable. So much of the territory beyond the barn and its adjacent pasture is new territory for the lambs, and they don't yet have the lay of the land.

Sheep who spend their lives on one farm develop a mental map of the farm,

though I doubt they visualize it the way we do, say on a GPS. I suspect their internal GPS consists of landmarks identified with sights, sounds, smells, and past experiences. They mentally flag these landmarks, which become their frame of reference. It's one of the reasons our adult sheep willingly move from one place to another: there will be something good when they arrive.

The lambs are entering uncharted territory, so all the sensory information is new. They do not know the routes, and they have no prior experience to guide them. They learn that the culvert that crosses the brook before the lane is a "no passing" zone. It's single file or take a dip in the brook. The lane is the turnpike that connects the lower farm with vast upper pastures we've been stockpiling for the lambs. Getting there is a trek. Crackerjack becomes their rudder. Still they take detours, exiting prematurely down the side of a ravine while the llama and lead lambs move ahead.

Flocking together and leaving the barn in unison is straightforward. Ini-

tially, with the path from the barn well worn by the traffic of many hooves, they follow single file. Then we have to cross a familiar pasture to get to the lane. This is a bit more challenging, as the lambs may mistake this field as their final destination. With Crackerjack on halter, I walk briskly while shaking a grain pan like a tambourine to build forward momentum. Mike or Holly brings up the rear to discourage dawdling. The lambs must keep sight of Crackerjack, their beacon, and the other lambs, since they don't know where we are going.

We traverse the pasture and tiptoe over the culvert. Then it's a hard right turn westward, up the lane. Stone walls along both sides serve as natural barriers and help direct the flow of traffic. There's a gate where the meadow portion of the lane reaches the wood and bends to the left—we've left it open in advance. The lane beyond the gate winds uphill through the woods. It's the main thoroughfare to the upper pasture. But if we're not careful, here's where things might fall apart. In addition to asking the lambs to move from the familiar to the unknown, we're asking them to do several things simultaneously that are counter to some very basic protocol for moving sheep. Sheep prefer moving from a dark area to a light area. The lane beyond the gate is tunnel-like beneath a canopy of scrub and mixed hardwoods. I anticipate that this will give them pause, but if we've gotten an early start or if I have the foresight to plan this outing for an overcast day, it's not a big deal. However, sheep like moving in straight lines, prefer moving downhill, and are suspicious about changes in the ground surface, but just beyond the gate the lane bends to the left and climbs steeply and the ground beneath their hooves goes from grass to gravel. If the flock isn't tight, or if some of the lambs are balky, the flock splits, with the slower lambs detouring into the sugar bush pasture on the right.

If the flock splits, the trek stops there. There's no sense in dividing the lamb flock. It's much simpler to reroute the llama and his close followers to join the AWOL lambs. The sugar bush pasture becomes a bivouac. We'll attempt the summit another day.

In a good year, when the lambs are focused and cohesive, we'll make it to the upper field in one jaunt. The hillside opens up into a field of mixed grasses, wild strawberry, bird's-foot trefoil, and clover. Thickets of wild raspberry and blackberry ring the field. In the center stands a lean-to made from rough-cut timber. It's surrounded by stone walls and birch and rogue apple trees. The lower limbs of cathedral pines provide more cover. We set up the water tank and a mineral feeder. It's a milestone—the young flock shifts its locus from barn to field. We'll visit twice a day, to ferry fresh water and do head counts. The lambs have the best grazing on the farm under the watchful eye of Crackerjack. Like pilgrims in a new land, they set up a colony.

Sheep will eat just about whatever's underfoot. It's both the easiest and most difficult factor influencing sheep management. In early spring when the grass is barely green, the sheep will nibble. When grass and weeds thrive, the sheep will browse contently. Before the snow buries the pasture for the winter, the sheep will graze the browned-out stubble. Sometimes they'll even scrape their way through a few inches of snow to eat what's underneath. This characteristic is thrifty and often gives new shepherds the idea that sheep are easy keepers because they will eat anything.

On the flip side comes the management responsibility. If left on a field too long, they will pick away at everything underfoot until there's nothing left, risking the health of both pasture and flock. Overgrazed pastures don't have time to regenerate. Flocks fed on overgrazed pastures are susceptible to internal parasites.

All sheep carry a level of worm load in their gut. It's neither unusual nor unhealthy for a flock's fecal sample to show evidence of strongyles. It is not a harbinger of doom but actually a normal fecal sample. Most vets will not advise drenching sheep against a mild showing of parasites.

Parasite management is an unsavory topic that no one, aside from savvy shepherds, wants to discuss. I had heard that sheep might get worms but honestly didn't spend a lot of time considering this when I got my first sheep. I was entirely engrossed in the process of naming them all, making friends, taking pictures, and hand spinning their glorious fleeces; I didn't consider the linkage between pasture management and flock health. Ultimately, good sheep management affects wool management. Ergo, good pasture stewardship is an essential but easily and often overlooked component of fiber farming.

In an overstocked pasture where the sheep are feeding on ground contaminated by droppings, they are reingesting parasite larva along with the grass. The healthiest head of sheep may never exhibit the most blatant symptoms of parasites, but the youngest and eldest flock members will tell the story—and be hit the hardest.

Raising sheep naturally on a hill in New England actually requires a fair amount of planning and human intervention, quite frankly at a time of year when I'd prefer to spend an afternoon gardening or knitting in the shade of my porch.

Rolls of portable fencing of poly twine threaded with an electrified wire make it all possible. We have miles of the stuff. One of our late-spring projects involves staking it out in the fields and then throughout the summer periodically rearranging it to keep the sheep grazing clean ground. In theory, you move the sheep onto ungrazed turf every five days or so. Keeping them on fresh grass in its prime vegetative state (of five to ten inches of lush growth) encourages the "good grasses." When the sheep are moved off a field, it's mown to crop any unpalatable weeds that the sheep have selectively worked their way around.

In practice, there's a lot more to it, and it's taken years to figure out what works for our sheep on our land. I'm always learning from talking to other shepherds about their grazing systems. Before owning my own farm, I took open fields for granted. In retrospect, it seems so silly. I'm not sure how I thought open land remained open. And even though I was surrounded by agriculture, the connection between farmers' fieldwork and the open vistas was not something I fully appreciated until my own initiation into the art of pasture farming and rotational grazing. Although I had set my sights on raising sheep, with wool as a principal crop, I quickly learned to focus on the underpinning of the entire enterprise: grass.

Not too far along in our farming tenure we realized that we had underestimated the impact of the previous owners' Holsteins on keeping our fields open. Within one cycle of seasons we could see the progression of milkweed, juniper, and multiflora rose. Many fields fenced with barbed wire were abloom with burdock, thistle, and nettle.

Left ungrazed, spring pasture quickly becomes a jungle of shoulder-high grass. Since the idea behind rotational grazing is to allow access to a week or so's worth of grass, then pick up the fence, reset it, and move the sheep to fresh ground, ideally, each section of pasture gets a three- or four-week resting period between grazing sessions. And since managing sheep means managing the internal parasites that go along with keeping sheep, moving them often and giving pastureland time to rest are crucial to the health of the land and of the flock.

THE KITCHEN GARDEN

IN THE SPRING, grass isn't the only crop on our minds. It's also time to start the kitchen garden. Unlike the manicured and landscaped beds around our house, the kitchen garden down at the lower farm is a free-form mosaic of herbs, perennials, veggies, and annuals. The plot started as a rectilinear bed measuring roughly sixteen by thirty feet running east–west. The back side of the studio shielded it from the north. The hay field bordered its southern edge. In my garden's first years a neighbor rototilled it each spring, and I would plant anew. Three years ago, my friend Ivy Palmer of Pitchfork Farm (who at the time was managing the Shelburne Falls Farmers Market) helped me reenvision my garden. A believer in no-till gardening, Ivy pitched the idea of planting in raised mounds.

That May, Ivy and I put spades to the dirt to create raised free-form beds following the contour of the yard in the space between the studio and the hay field. The garden's footprint now resembles an elongated arch. At the east end, three semicircular rows form a rough arc. A clematis-shrouded stick pyramid anchors the radials. Bright orange and red nasturtiums ruffle the base of the towering purple-flowered vine. The arcs are planted in a colorful slapdash of flowers (purple cone flower, coreopsis, Persian mums, salvia, anise hyssop, Russian sage, rudbeckia, bee balm, bachelor's button) and herbs (at least three varieties of basil, fennel, dill, oregano, chives). Thyme rims the stepping-stones.

The bed that runs along behind the studio forms the backbone for the rest of the garden. It's where I plant sunflowers, tomatoes, dahlias, and anything that grows tall and needs serious staking. Perpendicular to the "spine" run a series of shorter rows, like ribs. The middle row is really three mounds, each topped with a large teepee made from felled birch tree saplings. The structures are for climbers and creepers: scarlet runner beans, cukes, gourds, squash, morning glories. Ivy set the sapling trunks into the ground at an angle, at least eighteen inches down. We lashed them together at the top of each tripod. The structures stand eight feet tall. We left the fine twiggy branches at the top, for vines and birds.

The remaining spaces in the "rib" beds are planted with veggies and flowers, with no hard-and-fast rules. The configuration changes each year. I heavily favor anything that can go from plant to mouth: Sun Gold cherry tomatoes; sugar

snap peas; beans in purple, green, and yellow. I leave plenty of room for successive plantings of salad greens. A row for root vegetables and onions. Brussels sprouts. For several years I planted ground cherries but have since stopped. The plants became shrub-like; mice congregated amid the fallen cherries and papery brown husks beneath the plants. Corn snakes followed the mice. Although harmless (and actually good for the garden), the large brown-and-tan-flecked snakes unnerved me when they slithered between my legs.

I want color everywhere, so I plant annuals in the mounds right alongside the veggies. Marigolds at the end of each row by the hay field to discourage deer and rabbits. Lavender wands rule the perimeter rows. Their fragrant oiled foliage annoys the deer.

Flattened cardboard boxes suppress weeds between rows—the boxes in which my yarn comes back from the fiber mill. We sprinkle the cardboard with golden straw, spreading it right up to the edges of the planting mounds.

Composted sheep manure enriches the soil; we fork it into the beds each spring. I appreciate spring's slow arrival to the hills, since lambing is well behind me by Memorial Day weekend, when tender plants can first go into the ground.

At the height of summer, Holly and I snack on snap peas and beans from the garden between chores. At the end of the season, when we're drowning in squash, tomatoes, and brussels sprouts, I veggie bomb my friend Margaret at her restaurant and all of my nongardening friends.

A SUMMER MEAL FROM THE KITCHEN GARDEN

Recipes by Margaret Fitzpatrick

This meal came together one morning after walking my garden with friend and chef Margaret Fitzpatrick. All of these dishes were designed using the freshest local ingredients in season—sourced from our own farm and our neighboring farms.

SALAD OF GARDEN GREENS AND NASTURTIUMS WITH MARINATED GREEN, YELLOW, AND PURPLE BEANS; BEETS; AND SNAP PEAS, WITH LEMON TARRAGON VINAIGRETTE

Any greens from your garden can be used for this salad: red leaf or green lettuce, mesclun greens, arugula, or young spinach, in any combination. As we all know, greens are ready for picking when they are ready and need to be eaten immediately at their peak.

Ingredients for the Salad

- Fresh-picked salad greens
- Red beets, sliced in quarters
- Yellow and purple string beans
- Sugar snap peas
- Nasturtiums

Ingredients for the Vinaigrette

¾ cup lemon juice, preferably fresh squeezed
2 shallots, chopped
2 teaspoons salt
¼ teaspoon black pepper

½ cup fresh tarragon, chopped
2½ cups extra-virgin olive oil

Wash the greens in sufficient water to float them above the bottom of the container. Lift the greens out of the water and drain.

Place the beets in a pot of water and bring to a boil. The size of the beet determines the cooking time. When a small knife inserts into the beet easily, they are cooked. Drain water and cool.

Snap the ends off your beans. I prefer to snap only the stem end and leave the tail. Bring a pot of water to a boil. Blanch the beans for only a second or two by dropping them into the boiling water, stirring, and immediately removing and putting in an ice bath or running under cold water. Shocking the beans stops the cooking and preserves their bright colors while tenderizing the outer skin. The same should be done for the sugar snap peas.

Pick nasturtiums and check for any hitchhikers that may have come along from the garden.

To make the dressing, place all the ingredients in a food processor except the olive oil. Grind the flavors together, then slowly drizzle in the oil until it's incorporated. Adjust the seasoning to taste.

Arrange the lettuce around the outside of a platter. Arrange the three colors of beans, sugar snap peas, and beets in the middle. Drizzle the Lemon Tarragon Vinaigrette over the arranged salad and decorate with the nasturtiums.

LAMB KEBABS WITH ANISE HYSSOP PESTO

Tender lamb kebabs make a wonderful summer menu, and the majority of the preparation can be completed well in advance to allow more time with your company. The sweet, mild licorice flavor of the anise hyssop, an easy-to-grow herb, complements the earthy nature of the lamb. We've taken the liberty of blending the Middle East preparation with an Italian twist. This menu serves eight guests with plenty to spare.

Ingredients for the Kebabs

One leg of lamb or 4 pounds cubed lamb meat
4 cups olive oil
1 cup red wine
2 tablespoons sea or kosher salt
½ teaspoon black pepper
8 cloves garlic, minced
¼ cup fresh oregano, minced
¼ cup fresh mint, minced

Ingredients for the Pesto

1 cup anise hyssop leaves
½ cup basil
¼ cup mint
2 cloves garlic
¾ cup pecans, toasted
½ cup parmesan cheese
¼ cup lemon juice
1 cup extra-virgin olive oil

Bone out the leg of lamb or get a leg already boned and trimmed. Cut the meat into one-inch cubes, removing any sinew but leaving a little fat.

Create a marinade out of the remaining ingredients by mixing them together. Toss the meat in the marinade and refrigerate for at least 4 hours, if not overnight.

Soak some ten-inch bamboo skewers in warm water. This will help to prevent the sticks from burning on the grill before the meat is cooked through. Skewer the marinated lamb and set on a platter ready for grilling.

To make the pesto, wash the anise hyssop, basil, and mint in a cold-water bath and drain well. Grind the herbs in a food processor. Add the remaining ingredients and process into a rough paste. Place the pesto in a serving bowl and cover with plastic wrap until the time of

service. Make sure that the plastic wrap lies directly in contact with the surface of the pesto. This will help prevent oxidation, which turns the lovely green color dull brown. The pesto can be made up to a day ahead if care is taken to cover well.

For grilling: Preheat your gas grill on high heat. If you are using charcoal, start the fire well in advance to allow time for the hot coals to form and the flames to die down. Place the skewered lamb kebabs on the grill with the ends of the skewers off the edge and away from the heat to help prevent burning. Depending on how much wind there is and how hot your grill is, the kebabs will take approximately 8 minutes to cook with your turning them occasionally. It is important to let the meat sear on each side to retain the juices within, so don't be overeager with rotation. Allow time for the kebabs to cook on each side, about 2 minutes. You can tell when the lamb is cooked by removing a kebab and separating two of the inner pieces of meat. If they are still pink but not raw in between, the kebabs are ready to be removed to rest for 5 minutes on a platter.

ITALIAN COUSCOUS (FREGOLA)

This recipe uses a much larger pasta than in the Middle Eastern version—almost the size of barley.

Ingredients

3 cups fregola
2 teaspoons salt
6 cups water
⅔ cup extra-virgin olive oil
¼ cup fresh lemon juice
2 tablespoons oregano, finely chopped
Salt and black pepper to taste

Cook the fregola until tender but not mushy—approximately 20 minutes in the salted boiling water. Drain any excess water. Whisk the remaining ingredients together and dress the pasta while the fregola is still warm.

ZUCCHINI CAPONATA

Caponata, basically an Italian relish, is traditionally made with eggplant, but we have substituted with whatever is ripe in the garden on the day of the lunch. Pine nuts and olives may also be added as desired.

Ingredients

¼ cup extra-virgin olive oil
1 white onion, diced small
1 tablespoon garlic, minced
2 tablespoons capers
4 to 6 zucchini (depending upon size), diced small
4 ripe plum tomatoes, diced small
½ cup basil, finely chopped
Salt and pepper to taste

Heat olive oil in sauté pan. Sauté onion until clear. Add the garlic and cook only until the aroma of garlic is released. Add the zucchini and cook for 3 minutes. Add tomato, capers, basil, and seasonings and cook for 3 more minutes. Cool and serve at room temperature.

MAPLE-SWEETENED PEACH TART WITH MOUNTAIN BLUEBERRY COMPOTE

There is nothing more delicious than a ripe peach in season. Childhood memories of my mother treating us to sun-warmed peaches off the kitchen windowsill still bring a smile to my lips. Peaches that were so juicy, she would only allow us to eat them at the picnic table in the yard, with a napkin in one hand and sweet juices running down the other, right up to the elbow! This recipe's sweetness relies on the natural ripeness of the fruit.

Ingredients for the Crust

2 cups flour
¼ teaspoon salt
¼ teaspoon nutmeg
½ teaspoon ground ginger
1 cup unsalted butter
2 tablespoons sugar
1 egg

Ingredients for the Filling

8 ripe peaches
½ cup maple syrup
1½ teaspoons gingerroot, fresh minced
3 tablespoon arrowroot
1 teaspoon lemon juice
Pinch salt
Vanilla ice cream for serving (optional)

Mountain Blueberry Compote

4 cups small mountain blueberries
¼ cup orange juice
1 teaspoon orange zest
¼ teaspoon salt
½ cup sugar

To make the compote, reserve one cup of blueberries and place all of the remaining ingredients in a small saucepan and bring to a boil. Cook for about 10 minutes, stirring often until the blueberries have burst and released their juices. Add the last cup of blueberries and continue to cook until the mixture coats the back of a spoon. Total cooking time should be about 15–20 minutes. Pour into a serving bowl and let cool. The compote can be made in advance and kept refrigerated for up to two weeks.

To make the crust, put flour, sugar, salt, nutmeg, and ginger in a food processor. Add the butter and pulse until the butter is cut into dry ingredients about the size of lentils. Add the egg and pulse until the dough begins to come together. Do not overmix, or dough will not be flakey.

Press into an eleven-inch tart pan and chill in the refrigerator.

While the dough is chilling, prepare the filling. Boil a pot of water. Cut a cross in the top and bottom of each peach. Drop peaches into boiling water for two minutes or until you begin to see the skin of the peach separating from the meat of the fruit. Shock peaches in ice water. Remove skin. Remove pits from peaches by cutting them in half, retaining all juice in the bowl along with the halved, pitted peaches. Add the remaining ingredients.

Arrange the peach halves on the bottom of the tart crust with cut side down. Pour the remaining liquid on top.

Bake at 350 degrees for 20 minutes. Rotate the pan and bake for another 20 minutes or until the crust is brown and the arrowroot has thickened the liquids. Cool. Remove the tart ring and serve with blueberry compote and ice cream.

LAVENDER HARVEST

LAVENDER IS PERHAPS the most useful herb in my garden. During the growing season the fragrant leaves deter the deer. Some varieties have culinary uses. When the buds are harvested and dried, they can be used in sachets to shield woolens from clothing moths.

Research into natural moth deterrents led me to nearby Stockbridge Farm in South Deerfield, Massachusetts. On a June afternoon, farmer and herb specialist John Warchol and I stoop to examine wand-like stalks of lavender, heavy with tiny buds in shades of dusty purple. In addition to its being lovely, lavender's versatility as an herb led me to want to try growing some varieties in my own garden. "These are ready to harvest," John explains, "when the clusters of flowers at the base of the spires are just beginning to open." That is when the oils of the lavender plant are most concentrated and pungent.

For nearly a decade John and Mary Ellen Warchol have been raising herbs for culinary and decorative purposes on the farm where John grew up, which was at one time operated as a dairy. Beds that once produced corn and traditional veggies are planted with row after row of lavender plants, a shimmering sea of purple above mounds of silver-green foliage. Stockbridge Farm produces a half dozen varieties of lavender plants and more than forty varieties of basil.

John offers me a lavender primer. The English varieties, Hidcote and Munstead, are the first to flower. These are the plants Mary Ellen will use to infuse oils and vinegars for the kitchen. At the end of the row is the Grosso lavender, an elegant long-stemmed French variety that Mary Ellen will plait to form decorative lavender wands or will dry to make lavender sachets. Holding a spire to my nose, I inhale the intensive lavender—the concentration of oils in the French varieties makes them a favorite for scenting body-care products.

Two rows over I spy a large mass of yet another shade of purple, a rare strain of Czech lavender from the family *Lavendula angustifolia,* which they started from seeds given to them by their plant supplier seven years ago. Also highly fragrant, the stalks give off a lavender scent with a note of camphor. John explains that this variety is often used for medicinal purposes.

Late June is the height of lavender season. John will walk among the rows snipping the bud-laden stalks before they go into full bloom.

In the adjacent rows, the basil plants are still in the early stages of growth. John explained that people tend to think there is only one type of basil, the broad-leaf variety used in making pesto and marinara sauce. As we stand in the basil patch, he hands me leaves to sniff. One variety has a wonderful hint of lemon (the leaves are even tinged with yellow). Another has the powerful smell of licorice. Yet another has a cinnamon scent. And there are more. Several varieties of green-leafed, purple-stalked Thai basil (the fast growers, he tells me) tower above compact globes of diminutively leafed globe basil, resembling minitopiary. Then there's a host of purple and lemon basils.

My visit to Stockbridge Farm encouraged me to establish my own lavender border along my perennial and vegetable garden. I planted the French variety Grosso and harvested the fragrant buds in late June to make excellent sachets for storing with precious hand knits to protect them from moths.

Moths love wool. The dark corners of bureau drawers, recesses of closets, or baskets of yarn stash displayed on a shelf are attractive incubators for the eggs of the common clothes moth. Upon hatching, the larvae rely on wool, the dirtier the better, as the staple of their diet. There's nothing more tragic than discovering a peppering of holes chewed through a cherished hand-knit sweater, turning your favorite cardigan to lace.

The scent of raw, unwashed fleece is the clothing moth's siren song, which makes a good case for not storing raw wool for any length of time. But even washed wool is at risk. And hand-knit sweaters, socks, and blankets stored in chests and closets are equally attractive. The best defense against infestation is making your wardrobe inhospitable to the winged devils in the first place. Clean your woolens before you store them (the smell of perspiration, body salts, and food add to the attractiveness), and use olfactory camouflage, such as lavender sachets. If you make felted wool objects (or collect them), inspect them periodically. Olive oil soap, often used in felt making, may attract clothing moths.

The best protection for woolens in any form is to store them in a way that prevents access. I have a giant chest freezer in the garage—a perfect airtight storage for fleece and roving. I defy any moth to attack my stash. And while a freezer in the garage could potentially hold a lot of socks and sweaters, it's not the most convenient wardrobe when you're hurrying to get ready for work in the morning. Creating lavender pillows to place in bureau drawers or on closet shelves is a more practical option. You can start with your own plants, but be sure to harvest the buds just before they open. Use an old window screen as a drying tray, putting it in a dry place out of direct sunlight in your shed or garage. If you're not a gardener, you can browse the farmers' market to find a local herb farm for purchasing buds.

LAVENDER PILLOWS

Near the end of June, harvested buds of fragrant French lavender, valued for its intense essential oils, can be used to create lavender pillows to set in bureau drawers and blanket chests. Place one square in each corner of the drawer or chest where you are storing woolens.

Materials (makes one pillow)

- One 6" square of cotton fabric (a great use for remnants from quilting projects)
- Thread
- One 4" square of loosely woven muslin or voile
- 3 tablespoons (approximately ½ ounce) dried lavender buds
- Embroidery floss

Tools

- Quilter's pen
- 6" quilting square for tracing fabric squares
- Iron
- Shears
- Sewing machine
- Embroidery or tapestry needle

Using a quilter's marking pen or pencil, measure and trace the lines for the ¾-inch border along all edges of the wrong side of the cotton fabric square. At the intersection of lines, mark the diagonal for the corners.

Fold the tip of each corner along the marked diagonal lines inward and press. Trim away the outermost tip at each corner. Fold the corner again and press.

Turn the sides in ⅛" and press. Then fold the edges in to form the mitered corners. Press carefully.

Insert the muslin (or voile) square into the "window" created by the cotton square, slipping it beneath the pressed edges. At the sewing machine, stitch around the square close to the inner fold line. Leave a one-inch opening for adding the lavender buds.

Using a spoon, paper tube, or small funnel, add 3 tablespoons of buds to fill the lavender pillow. Back at the sewing machine, stitch the opening closed.

A pair of criss-crossed stitches of embroidery floss creates the decorative tuck in the center of the pillow. This step could be eliminated and the sachets would work equally well.

Cut two 6" lengths of floss. Using a sharp embroidery needle, make a stitch into the voile face of the pillow, just a tiny bit off-center. Pull floss through to the back side, and make a ¼" diagonal stitch into the back fabric of the pillow, coming out on the front. Repeat with the second piece of floss. The two stitches should form an X on the back side of the pillow. The tails are on the front side (see illustration). Gather the four strands of floss and create an overhand knot. Trim the strands to desired length to create a tuft in the center of the pillow's face.

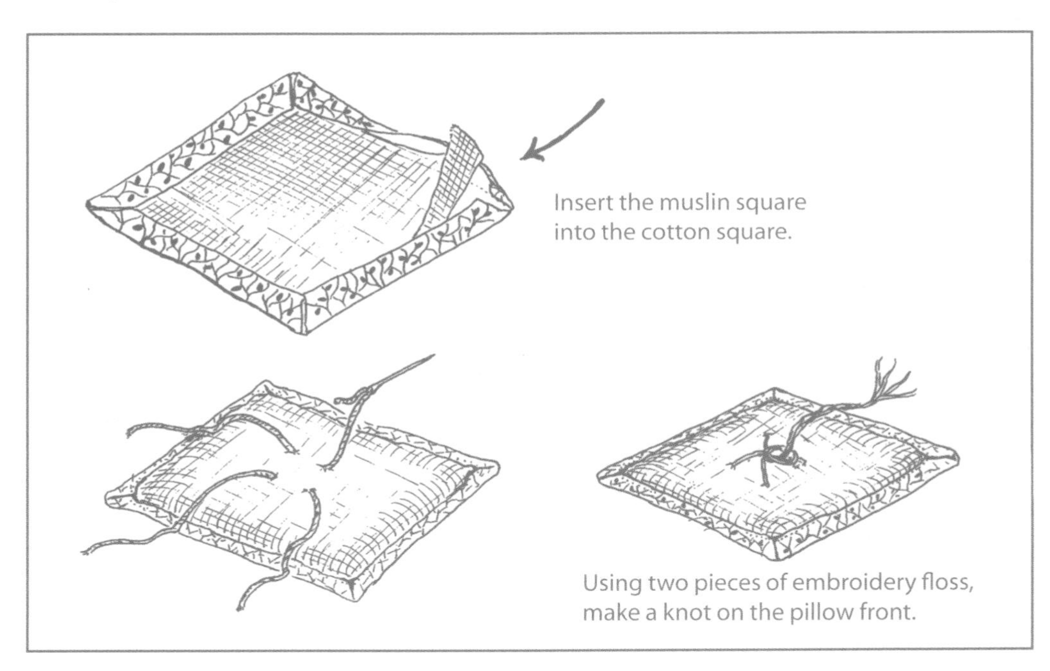

Insert the muslin square into the cotton square.

Using two pieces of embroidery floss, make a knot on the pillow front.

BRIDGE OF FLOWERS CARDIGAN

Designed by Lisa Lloyd

Come midsummer, the farm and the whole area are abloom with flowers. Not far from Springdelle Farm is Shelburne Falls' most famous garden, the Bridge of Flowers. The garden spans the Deerfield River in the heart of the village, just above the glacial potholes of Salmon Falls. Originally constructed in 1908 as a trolley bridge linking the towns of Buckland and Shelburne, its conversion from trolley bridge to pedestrian footpath and garden occurred in 1928 after trolley service had stopped. The bridge became the project of the Shelburne Falls Women's Club, a volunteer organization that to this day oversees the maintenance of the bridge gardens. It is one of our favorite strolls on a summer evening.

This cardigan perfectly accompanies a camisole, tank, or strappy dress and can be dressed up or down for cool summer evenings. The garment is knit in one piece from side to side, beginning with the left sleeve. Once you have bound off, sew the sleeve and side seams together, add a button, and you're finished.

The yarn for this sweater is a blend of my own devising, consisting of Cormo wool (70%), alpaca sourced locally (20%), and Bombyx silk (10%). This fiber combo gives the garment amazing drape, softness, and sheen. Spun semi-worsted style at the mill, the fibers have been combed for smoothness, eliminating the itch factor. It's what makes this a great yarn for summer garments worn next to the skin. The DK gauge is cozy yet lightweight. The high percentage of bouncy Cormo wool ensures this garment will not lose its shape.

Finished Measurements

- Chest: 34 (38, 42, 46)" (86 [96.5, 106.5, 117] cm)
- Length from shoulder: 16 (17, 17, 18)" (41 [43, 43, 46] cm)

Yarn

- Foxfire Fiber & Designs Cormo Silk Alpaca (70% Cormo, 20% alpaca, 10% silk); 190 yd [175 m]): 5 (6, 6, 7) skeins. Shown in Autumn Hydrangea.

Needles

One 24" (60 cm) or longer circular needle size US 5 (3.75 mm)

Notions

- Stitch markers
- Stitch holder
- Tapestry needle

Gauge

20 sts and 32 rows = 4" (10 cm) in Body Pattern. Adjust needle size as necessary to obtain gauge.

Border Pattern

Rows 1–5: Knit.

Row 6: Purl.

Row 7: K1, *yo, k2tog, rep from *, end k1.

Row 8: Purl.

Rows 9–16: Repeat rows 1–8.

Rows 17–20: Knit.

Body Pattern

Row 1 (RS): Knit.

Row 2: Purl.

Row 3: *K1, p1, repeat from * until 1 st rem, end k1.

Row 4: Purl.

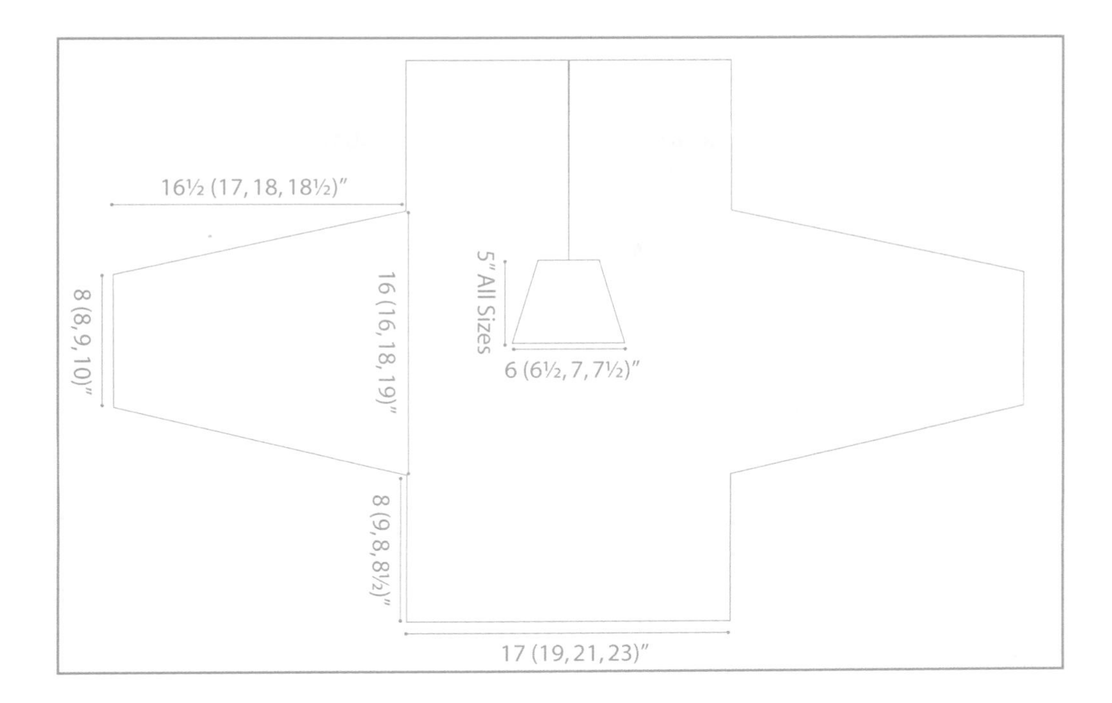

Note: Cardigan is knit in one piece, side to side, beginning with Left Sleeve.

LEFT SLEEVE

CO 40 (40, 46, 50) sts. Work rows 1–20 of Border Pattern. Work row 1 of Body Pattern. Continue working Body Pattern and at the same time inc 1 st each side every 4th row (WS) 8 (6, 8, 9) times, then every 6th row 12 (14, 14, 14) times, incorporating new sts into pattern as follows:

P1, M1R, work across until 1 st rem, M1L, p1—80 (80, 90, 96) sts. Work even until sleeve measures 16½ (17, 17½, 18)" (42 [43, 44.5, 45.5] cm) from beg, ending with a WS row. (Note the rows worked after the increases are complete are to be used for right sleeve.) Place sts on a stitch holder. Cut yarn.

LEFT FRONT AND BACK

CO 40 (46, 40, 42) sts for Back and work across these sts in Body Pattern, work across 40 (40, 45, 48) sts of left sleeve in patt as established, pm, continue working across remaining 40 (40, 45, 48) sts of sleeve, then CO 40 (46, 40, 42) sts

for Left Front and work in Body Pattern, joining to sleeve and Back sts—160 (172, 170, 180) sts. Work even in patt until body measures 5½ (6¼, 7, 7¾)" (14 [16, 18, 20] cm) from top of sleeve, ending with Row 4 (2, 4, 2) of Body Pattern.

DIVIDE BACK AND LEFT FRONT AND SET UP NECK SHAPING

Note: To keep continuity of Border Pattern during neck shaping, work Row 7 of Border Pattern as follows: If there are an odd number of stitches to work, begin with knit 1, then work Row 7 of Border Pattern. If there is an even number of stitches, work Row 7 as written. This also applies for Right Front neck shaping.

Work across Back sts to m, remove m; with second ball of yarn BO 3 sts (Left Front neck edge), work to end of row for Left Front. (Back and Left Front are now worked with separate balls of yarn.)

Next row (WS): Work to 3 sts before neck edge, p2tog tbl, p1; work across Back sts as established to end.

Decrease Row 1 (RS): Work across Back sts in Body Pattern; at Front Neck edge, BO 3 sts and work across remaining Left Front sts in Body Pattern for 4 (6, 8, 10) rows of neck shaping as established, then begin working Border Pattern on Front sts only for remainder of Front. (See note above for continuity of Border Pattern.)

Decrease Row 2: Work Left Front in patt as established until 3 sts rem, working k2tog, p1, on WS rows that are knit, and p2tog, p1 on WS rows that are purled; work across Back sts in Body Pattern.

Repeat Decrease Rows 1–2 four more times.

Repeat Decrease Row 1, but dec 4 sts at Front neck edge (instead of 3).

Repeat Decrease Row 2.

55 (61, 60, 65) sts remain for Left Front. Work 12 (14, 16, 18) rows even—20 rows of Border Pattern are complete.

Next row (RS): BO all Left Front sts. Cut yarn and pull end through last loop.

RIGHT FRONT

With a second ball of yarn, CO 55 (61, 60, 65) sts for Right Front onto left needle, leaving Back sts on right needle. Turn (WS) and work Border Pattern over Right Front sts, beginning with Row 2 of Border Pattern. Work across Back sts as established. Cont to work Back and Right Front separately. Work a total of 11 (13, 15, 17) rows, ending with a WS row.

SHAPE RIGHT FRONT NECK

Work across Back sts as established; at Right Front neck edge, CO 4 sts and work to end in established Border Pattern.

Increase Row 1 (WS): Work across Right Front sts until 1 st rem, M1, purl 1; work across Back sts. Incorporate new sts into Border Pattern.

Increase Row 2 (RS): Work across Back sts as established; at Right Front neck edge, CO 3 sts and work to end in established Border Pattern.

Repeat Increase Rows 1–2 for a total of 6 times, and at the same time, when Border Pattern is complete (20 rows), work in Body Pattern over Right Front for 4 (6, 8, 10) rows; keeping Back sts as established, continue to work Right Front and Back separately.

Note: Be sure to keep continuity of Body Pattern during neck shaping on Right Front. Once shaping is complete, there are 80 (86, 85, 90) sts for Right Front, ending with a WS row.

JOINING THE BODY

Next Row (RS): With one ball of yarn, work across all sts for Right Front and Back in established patt. Work even for 5½ (6¼ , 7, 7¾)" (14 [16, 18, 20] cm), ending with a WS row.

BO 40 (46, 40, 42) sts at the beginning of the next two rows—80 (80, 90, 96) sts.

RIGHT SLEEVE

Note: If additional rows were worked (and noted) to obtain length of Left Sleeve, work the same number of rows here before beginning decreases.

Decrease Row (WS): P1, p2tog, work across until 3 sts rem, p2tog tbl, p1.

Work Decrease Row every 6th row 11 (13, 13, 13) more times, then every 4th row 8 (6, 8, 9) times—40 (40, 46, 50) sts. Work Border Patt for 20 rows. BO loosely purlwise. Cut yarn and pull end through last loop.

FINISHING

Sew sleeve and side seams.

LOWER BODY EDGING

With RS facing and beginning at Left Front lower edge, pick up and knit 170 (190, 210, 230) sts evenly around lower edge of body, ending at Right Front lower edge. Knit all sts for 3 rows, ending with WS row. BO loosely purlwise. Cut yarn and pull end through last loop.

FRONT AND NECK EDGING

With RS facing and beginning at Right Front lower edge, pick up and knit 60 (65, 65, 70) sts evenly along Right Front edge to Neck, pm, pick up and knit 122 (130, 138, 146) sts evenly around neck edge, pm, pick up and knit 60 (65, 65, 70) sts evenly along left front edge, ending at lower edge—242 (260, 268, 286) sts.

Next Row (WS): Knit until 1 st rem before the first m; knit into the front and back of the next st, sl m, knit into the front and back of the next 2 sts, work until 2 sts before second marker, knit into the front and back of the next 2 sts, sl m, knit into the front and back of the next st, knit to end.

Next Row: BO 61 (66, 66, 71) sts of Right Front, ending at m, remove m, knit 1. Turn and knit 2. Knit the 2 sts again, turn, knit the 2 sts (all sts should be on the left needle and button loop formed); BO all sts. Cut yarn and pull end through last loop.

Sew button opposite loop; weave in all loose ends.

Wash and block gently to measurements.

Hard Work NEVER
Hurt Anyone

FUN WITH TRACTORS

When we first started farming, Mike and I had rather vague notions of what field work entailed. We assumed that haying would take care of the hay fields and the sheep would take care of the rest.

After wading into our enterprise we realized that when it comes to fiber farming, field work requires a fair amount of sweat and manual intervention—often in the form of hand pulling thistle and burdock plants and lopping down multiflora rose. Our pastures were well sown in all three. Spiny globes of thistle seed heads and prickly bracts of burdock easily became embedded in the fleeces, and so they became my nemeses. The hand-to-weed combat approach worked okay for isolated patches—a stalk of burdock by the gate, a stand of thistle on a hillside. But fields overrun with invasive plants called for more aggressive measures.

The extent of my experience with outdoor power equipment at the time was limited to using a lawn mower and a handheld string trimmer—the tools I arrived with to start a farm. As it became clear that a handheld trimmer wasn't exactly the right tool for hacking through tough stalks of burdock or clipping swaths of milkweed in the middle of the field, I became a regular at the local power equipment dealerships and explored the options. A noisy self-propelled walk-behind string trimmer with heavy gauge cutting cord was a step in the right direction. It worked well for jobs like clearing a run so I could set up electric fencing. But mowing a five-acre field with a walk-behind string trimmer is not the way to go. The wheels got hung up in dips, and it took a fair amount of muscle to guide the machine through uneven terrain. The weeds continued to have the upper hand until I discovered the beauty of the brush hog.

A brush hog is a six-foot-wide steel mowing deck that gets hitched behind the tractor and powered by a PTO. The steel deck houses two gigantic whirling

mowing blades. After studying the manual and taking many notes, I practiced on the flat.

My first time out I used it to clean up the edge of a ravine the sheep had been grazing. I've always been a bit equipment phobic, shying away from noisy, large machines with sharp moving parts. A mechanically challenged Luddite, I still cannot hook up the brush hog without help. No matter how many times Norm Davenport has shown me how to do it, it always seems to require some fiddling around, which completely stumps me, since I have no idea what does what. If one must fiddle, one had better know what to fiddle with. It still escapes me, so I leave the fiddling to Norm.

But the brush hog was still another revelation. Engaging the mower was straightforward: lower the deck to the ground, engage the rear PTO, and open up the throttle. Empowered and unstoppable, I set out to tackle bigger projects. The lower half of the pasture behind the open barn, which is how we refer to the free-stall barn, had gone reedy. For two summers it had been far too wet to wade through with the string trimmer, but in this dry year it looked like a good first project for my brush hog. I asked my neighbor Bob what he thought.

Scratching his chin, he said, "Well, I suppose you could. Might hit a few nuggets."

By nuggets Bob meant rocks. The fields are studded with them. I knew where most of the large ones lay in the upper, dry reaches that I'd been maintaining with the string trimmer. But the low marshy area had grown up a bit. It was uncharted waters. Anything could lie under that grass. Emboldened by my passes around the edges of the hay field, I was game.

I set out to make a first circle of the perimeter. By keeping the bucket on the front of the tractor just inches off the ground, I figured I'd get wind of any rocks ahead by catching them with the leading edge of the bucket before actually mowing over them with my blades. I went slowly on my first pass, with my hand on the PTO switch just in case, keeping an eye on the fence line and mowing as closely as I dared without actually mowing down my woven-wire fence. With a clean first pass edging the field, I started a second circle, keeping the mower parallel to the first pass. I probably went a little faster as my confidence grew and I got a feel for how much of this heavy wet grass the mower could chew through without bogging down. In the thickest parts, I pushed the throttle higher to keep the rpms up. And that's when I bit my first nugget. I am telling you there is nothing more jarring than two steel blades clipping the top of a boulder. The sound is ungodly, the vibration of steel grinding stone enough to knock the fillings out of your teeth. In a panic I shut down the PTO and stopped in my tracks. Lifting the deck, I pulled forward to assess the damage. The boulder lay hidden like an iceberg amid the grassy hummocks. The blades had ground its tip right off. Peering beneath the deck, I could see that my blades looked like they'd been dealt a blow by a sledgehammer. On legs that felt like water, I walked the rest of the entire pass, poking the earth ahead

of me, checking for nuggets. My serenity shot for the rest of the day, I circled the field first on foot before each successive pass with the brush hog, giving every rock I found a wide berth.

I still hit plenty of rocks with my brush hog. In fact, I've come to accept that mowing rocks is part of mowing the fields when I clean up after the sheep. In most fields, I know where the big ones lie and can avoid trauma to my blades by steering clear. And then there are the ones I forget about each year until I hit them, like old friends. I have nicknames for some of them: Matterhorn, Pebble, Nugget.

Once the lambs have shifted grazing grounds in early summer, I mow the gnarly field on the downhill side of the birthing barn. It took years before I worked up the courage to do this. For one thing, it's steep. One of the first lessons my neighbor taught me about operating equipment on a grade was to always drive up- and downhill, never across. It reduces the chance of hitting something and tipping the entire rig. This field holds its own special challenges. The top is a bit of a minefield. The remains of a dismantled barbed wire fence lie hidden in the undergrowth, something I discovered when I ran over a half-rotten fence post a year ago. I was able to power down the PTO before the rusted strands of barbed wire got snared in the blades. Halfway down the hillside, thorny tentacles of multiflora rose camouflage a sea of rocks. Several drainage pipes come to daylight about midway down. I've memorized their locations and maneuver around them.

But the most treacherous part of that field is at the toe of the hill, where the ground goes from firm to Jell-O in a matter of a few feet. The farm is called Springdelle for the many springs that bubble up to the surface. I discovered this one the hard way the first year I drove straight to the bottom of the hill and attempted to turn left and ended up sunk to my chassis in mud. It wasn't just wet ground, it was a tar pit. My tires spun, but the ground was like pudding, and there was just no traction to be had. Fortunately, the ground just a little above the hole was good. Norm was able to haul me out using his much bigger tractor and a set of chains, but not before I'd made one whale of a rut in the bottom of the field.

I wish I could say that I'd learned from that and it was an aberration, never to happen again in that spot. But it's not true. In fact, I get caught in nearly the same place in that field almost every year. Depending on how wet a spring, the Mason-Dixon line separating field from swamp migrates. In drier years I can get fairly close to the scene of the first mishap (though I have never ever gone quite so far down as my initial foray). In other years, "the wet" starts higher uphill. I've learned to watch for signs in the vegetation, steering clear of sedge and bracken and watching the big rear tires of my tractor. If the wheels come up shiny, I know I'm bound for trouble.

HAY DAY

On the back side of summer, we again wait anxiously for a clear window of days for making hay. Second-cut hay gets put up in forty-pound "square bales" (which are actually rectangular in shape) and stacked in our hayloft. Cut grass must be crisp and dry before baling, though hot, late-summer days tend to be thick and hazy. When we're pushing it into September, the dew doesn't dry until midmorning. And the hill's shadows creep across the field by three in the afternoon. Popcorn thunderstorms are not uncommon. The later in the season, the narrower the window for hay harvest. Again, I get anxious. The dance becomes more precarious.

Making good squares is a lot of time and work. I often ponder how to convert my barns and feeders to feed round bales to my flock. But round bales are meant to be placed intact inside a giant round-bale feeder for cows to eat at will. I experimented with it one winter when square bales were scarce. My sheep ate themselves silly, got fat, and wasted a lot of hay by pulling it out of the feeder and spreading it all over the barn floor. What didn't end up on the floor got lodged in their fleeces, ruining the neck wool and top lines of almost

every fleece—I was very upset on shearing day, looking at all the chaff embedded in the wool. We also had lots of lambing trouble that spring; I had to pull a lot of really big lambs from seriously overweight ewes.

My barn and hayloft were designed for storing and feeding traditional square bales. On the uphill side of the barn, the loft is accessed by seven sliding doors. We can back in the hay wagons to off-load and usually start by filling the south end of the barn, working our way north. At the north gable end of the barn, we use a motorized conveyor to carry the bales up into the loft, where a couple sets of hands stack them neatly.

A thrumming hay mower is a welcome sound on a clear September morning—prime time for second cutting. Every farm on the hill is making hay. If Norm's tractor pauses for a moment, I can hear the distant purr of diesel from the neighbor's farm on the other side of the hill. A look at the sky will also tell you which farms are mowing. The sound of a haybine is a dinner bell for the

turkey vultures. They arrive in twos and threes and glide in figure eights above the mowing, targeting the unfortunate snakes and field mice that get maimed by the mower.

Compared with first mowing, the second harvest is smaller in volume, but there is also more leaf than stalk, making hay that is more packed with nutrients. It's the right feed for small ruminants. The potpourri of leaf and legume from our best fields will feed the gestating ewes in winter.

Norm's mower is actually a mower and a conditioner. While cutting the grass, the machine "beats up" the grass. A series of stripper fingers breaks up the waxy coating of the stems as they are cut, so they will dry faster.

Fresh-mown hay is lush and damp. It needs good weather to dry, and square bales need to be crackling dry. Any moisture causes the bales to mold and poses the potential risk of fire hazard in the barn.

Drying hay is a process of shake and bake. Norm or Lisa or Fred runs the tedder, a device with gyrating tines that lifts the cut grass and spreads it out to dry on the field. Then they come along with the hay rake, which grooms the fields into tidy windrows to dry. Ted, rake, dry. Rake, dry. Rake, dry. Each raking flips the hay, exposing more of it to the airflow. Drying takes about two days. We hope for a dry breeze to speed the process.

You can actually see the grass drying. The first windrows on a newly mown field are yellow-green. By the second day of flipping and raking, the grass takes on a bluish-green hue. As it loses moisture, it takes more of a slate-blue-green cast, and you know the sun and wind have done their work. I'll walk through the windrows, testing handfuls by twisting the blades in one direction and then in the opposite. If the grass is crackling dry, it should pull right apart. And then you know it's time to bale.

The hardest part of the job is rounding up enough hands to take in the hay bales. When the Davenport kids were younger, we could count on a gang of them arriving with friends to pitch in. Now that they're grown and working jobs of their own, we beat the bushes looking for anyone willing to help out. For some reason helping with hauling hay does not have the same allure as helping with the spring shear. It's hard to make it enticing. Friends show up once for the novelty of it, but unlike shearing day, the novelty of hay day quickly wears off.

It's late August and we've had a bear of tropical humidity for weeks. It's sticky but not raining, so we're making hay. Norm starts the baler just after lunch. We've got two fields raked and ready, but the weather report has called for a chance of late-day shower activity, which has upped the ante. All the grass needs to come in within a matter of hours. While Norm's running the baler, I make a few frantic calls to round up some extra hands. Gale stops by to pick veggies in the garden and then offers to stay and help.

Norm tows the square baler around the field, an old John Deere that he's been telling me is on its last legs since we started working together seven years ago. The grass is sticky, causing bales to jam up inside the machine. He keeps hopping off the tractor to check the weight on the bales and tinker. Then back on the machine. You can hear the distinctive sound of Norm's square baler anywhere on the hill. *Ka-chunk, ka-chunk, ka-chunk.*

Norm tells me this baler, a John Deere circa 1969, developed its signature

"thunk" when Norm's dad was running the baler on its maiden voyage o'er the windrows, where along with the cut hay it forked, scooped, and baled a round-toed shovel that had inadvertently been left in the field. Chunks of shovel stopped the machine cold—after it had neatly folded them, handle, wooden shaft, and steel spade, right into a bale of hay. Four decades later, this baler is still in operation. Its erratic thunk is the only reminder of its tangle long ago with the shovel.

Norm tows a buncher attached to the baler—a large open sled with steel runners along the bottom and a plywood door at the back end. The buncher collects and holds on to the bales as the baler spits them out. Once the buncher is full, Norm pulls a string to release the bales all in one spot. It saves lots of steps when it comes to collecting. Fred is driving the tractor with the hay wagon. Mike, Holly, and Brian, Fred's younger brother, have already started stacking the first load.

Picking up two fields in one afternoon takes some hustling. One or two people stand on the deck of the wagon. If you're the stacker, everyone is pitching bales your way onto the wagon. You lift and stack them, starting at the front of the

wagon and working your way back. There's a pattern to stacking hay, in order to keep the load stable and fit as many bales as possible onto each load. The wagons hold about 100 forty-pound bales. As soon as we've filled a wagon, a team follows it back to the barn. As stacker, I often end up on top of the load, pulling up the last of the bales. So I catch a break, riding on top of the load on the way to the barn. Which means I get to toss bales down to everyone else once we've backed up to the sliding door in the hayloft. We take turns throwing bales at each other all afternoon, load after load. Occasionally, the baler breaks a shear pin. *Ka-chunk, ka-chunk, ka-chink.* If we've caught up with gathering, everyone gets to hang out and eat some granola while Norm fiddles with the baler. Then it's back to work.

When we can round up enough hands, it's possible to have two teams gathering bales. The wagons pass each other on the way to and from the barn. The work is hot and sticky and dusty. Although I'm hot and my shirt is soaked with sweat, I've learned it's not worth working in short sleeves. The prickly ends of cut hay are needlelike, etching nasty red crosshatches onto my forearms. We suck down water from gallon jugs on the wagon.

Norm and I keep count. There's a counter on the baler that keeps track of

the number of bales it punches out. But sometimes bales break or we get "sinkers," heavy, wet bales from the outer windrows that will not keep. Rather than our stacking them with the dry hay in the barn, they get fed right away to the Davenports' cows. I count as we unload and stack at the barn. Between the two of us, we have a fairly accurate tally of the yield.

By late afternoon, we've finished one field and are working double-time to gather the bales in the field behind the studio. It's the field closest to the barn, which is a good thing. The sky has turned an ugly shade of puce, and the air thickens. Gale and I are hauling away, side by side. I look at her and laugh—her blond hair is standing straight up on end, like a dandelion halo. She can't see it, but it's really funny, and everyone stops to take a look. Then I take off my hat and we all laugh because where my hair isn't matted down with sweat, it stands up on end too. And then we hear the first clap of thunder. No one laughs. We hustle like hell to get the remaining bales onto the wagon. The wind picks up, sweeping in from the south. I can hear the train whistle clearly. Though the tracks are several miles away, closer to the river, the sound of the train means time for rain.

The sky lets loose as we're unloading one of the last wagons. There's not time to unload one wagon, so we push it into the carriage barn for cover until the storm passes. The Davenports scramble to park tractors in buildings. We're done for the day, and the hay is under roof. We'll finish unloading in the morning.

Each hay season plays out differently. I fantasize about the year we'll have second cutting in the barn by Labor Day and then kick back and toss back a few beers, knowing the flock will be well fed in winter months. But this is usually not how it goes. In my tenure at Springdelle Farm, there have been only two years when this has been the case.

Relief doesn't sufficiently describe the way I feel about finally having all of my barns stashed for winter. In years of wetter-than-wet summers and fickle autumn weather, I worry about having enough winter forage for the sheep. To go shopping for hay would be unimaginably expensive for a flock the size of mine. Farming sustainably means balancing the size of a herd with the capacity of the land to feed the herd—in all seasons. I'm committed to keeping a flock size that balances the carrying capacity of our land, but that changes from year to year

FALL

Fall is the time to wrap up summer's business in the wake of Autumn's equinox—getting hay into the barn, turning out the flock to clean up hay fields, and pulling late-season veggies from the garden.

In September the milkweed meadows are festooned with monarch butterflies. We wait for them to migrate before brush hogging the fields. By early October the bluebirds have flocked up and vacated the farm. I note a skyful of swallows late one afternoon as I unlatch the barn gate; they swoop and converge before plotting their seasonal retreat. Arrows of Canada geese flap noisily southward.

Fall is also a time of departures. Fleece that flew from sheep to skirting table to spinning mill in spring comes home to roost, briefly, transformed into skeins. The skeins take a detour through my studio for dyeing, and then, fully fledged, fly into the hands of knitters at fall fiber festivals.

We celebrate color—in the yarn and on the hills and in the dye pots. Golden leaves carpet the lane leading to the upper pasture. And then the season takes a turn and fall becomes all about preparing for winter and thinking ahead to the next year.

AT THE MILL

In September we summon the flock back to the barns group by group for general maintenance. I wade hip deep through a holding pen of ewes, assessing their body condition by working my hand beneath their fleece above their pelvis. I make note of who should be crutched out (cleaned up around the butt with hand shears) before breeding. The month's to-do list includes trimming the rams' hooves and taking a hard look at which lambs will remain and which will go. In the midst of administering to sheep, in the studio I am knee-deep in boxes of yarn and roving returned from the spinnery. The fall is the time when the work of spring is realized as skeins of yarn ready for washing, dyeing, and tagging. No matter where I turn, there's wool to be dealt with in one form or another.

Of course this process started months ago in spring, on the heels of shearing and skirting, when I entrusted my wool clip, the efforts of an entire year, to the yarnsmiths at the fiber mills. When I first investigated having my own wool mill-spun more than a decade ago, there weren't as many options as exist today. Back then, most mills in the Northeast and elsewhere had huge minimum

quantities. I would have had to accumulate several years' worth of wool clips to have enough for a single yarn run. The increase in small fiber farms in the past decade has fostered a niche industry—the small spinnery. Micro-spinneries (sometimes called "minimills," although "Mini Mill" is actually the trade name for a company that produces a line of cottage-industry fiber-spinning equipment) are the microbreweries of the fiber world. These small independent businesses bridge the gap in processing, making it possible for small-scale fiber farmers to create a value-added product. The symbiotic relationship between fiber mill and fiber farmer has resulted in a broader than ever range of yarn and fibers for knitters and hand spinners.

The gestation of lambs is about 145 days—the span of time it takes to transform a twinkle in the ram's eye to a woolly lamb bounding through the gate. The gestation of yarn has many variables and is a bit harder to predict. For me the yarn countdown begins on the heels of shearing day, the moment I ship boxes of raw skirted fleece to the fiber mill for transformation into finished

skeins. It is a very stressful moment. So much depends on the outcome of fiber processing after a year of careful fiber cultivation. Once those pungent boxes of wool go out the door in late spring, my work with this particular batch of fiber is done until many weeks later when smaller, tidy boxes of skeins return from the mill.

When it comes to making yarn, there is no one recipe that works well for all varieties of wool. And that is one of the reasons that communication between yarn farmer and yarnsmith is so important. Not all fibers can be all things yarn. The most lovely Border Leicester fleece may never feel quite soft enough to knit a bed jacket, and the most sound Cormo fleece will not hold up without reinforcement in a sock. Sheep breed, age, quality of fleece and how it was raised, and method of spinning are all determining factors in the outcome. When it comes to selecting a spinnery, yarn farmers may travel one of two main roads: the woolen route or the path of semiworsted.

My background as a hand spinner served as a useful guide in understanding

these two fundamental differences in yarn spinning styles. I was taught to spin worsted style, though at the time I had not a clue there was more than one way to do it. Short drafting small amounts of fiber at a time and putting a fairly high degree of twist into the fiber produced yarns that were strong, smooth, and sometimes even wiry, depending on the fiber I used. Although it made very good warp yarn for weaving, it was not lofty, and items knit from my very first yarns were not particularly soft. But I kept spinning them just like that because I was thrilled to be able to make yarn and didn't know any better.

It wasn't until I began reading, talking to other spinners, and taking workshops that I realized it was possible to expand my repertoire as a hand spinner. I remember admiring a hand-spun, hand-knit hat in a gallery exhibition at a fiber arts retreat, noting how the yarn felt and behaved so differently from the yarns I had been spinning, even though it was produced from a very similar type of wool. That observation opened the window to the world of woolen spinning. I had heard the terms *worsted* and *woolen* but really didn't understand or appreciate the differences until I sat down to master making them by hand. Where worsted yarns are spun inchworm style in an effort to keep the fibers as smooth and aligned as possible, woolen spinning works a twist into an airy, jumbled mass of unspun wool while not losing the airy, jumbled quality of the fibers. I suddenly found it was possible to spin lofty, fluffy yarn from my own sheep. And it was an epiphany. It gave me a great deal of insight into the various routes for processing at the fiber mills. Suddenly I understood why different varieties of yarns were available.

When I hankered for a lofty single-ply yarn from the wool of my Border Leicester hoggets (yearling ewes) combined with mohair, I consulted David Richie at Green Mountain Spinnery in Putney, Vermont.

Using vintage mill equipment, David and the spinnery's team of yarn artisans work closely with local fiber farmers both in the production of the mill's own line of woolen yarns and in the creation of custom-spun farm yarns. In the wool shed David opened my boxes for a hands-on look at my fibers. Knowing that woolen carders are limited in terms of the length of fibers they can handle without the wool's getting hung up on the drums, I had put my Leicesters on a shearing rotation of twice in fifteen months. It kept the fiber length under five inches. Although David felt the length would work, the slipperiness of the fiber gave him pause. Both mohair and Leicester wool are glossy fibers. When I asked him why that was a problem, he told me you can't know for sure that it is until you card the fibers and discover that the "web," the gossamer curtain of carded fibers, simply falls apart. Following his very strong hunch that with this initial batch of fibers we could have an unhappy outcome, I brought another hundred pounds of finer, softer wool from my crossbred sheep. This is closer to the type of wool for which the woolen carding equipment was built. The percentage of fine to slippery fiber gave us the confidence to go ahead and run with the yarn. And my Upland Wool and Mohair yarn was born. Even with my hand-spinning background, I wouldn't have known about the slippery-to-fine ratio issue without David's expertise. He knew his equipment and he understood fiber behavior. I relied on his judgment to transform my hogget fleeces into delightfully lofty woolen skeins.

Green Mountain Spinnery scours all fibers in-house, using methods that are gentle on both the fiber and the environment. On scouring days the air in the washroom is humid and smells like my barnyard on a steamy July afternoon. Against the wall, raw fibers steep in a large basin. Then they go through a series of gentle machine washes. Environmentally friendly cleaning agents release grease and grit from fiber. The water is recycled, once the lanolin is removed, to reduce consumption. The fibers are transferred to a spinning drum to shake out excess water, which reduces drying time in the large industrial dryer.

The mill's behemoth Davis and Furber wool card occupies the entire west end of the building. Washed locks of wool fed into the large hopper on one end of the carder travel through a series of drums and cylinders lined with bristling stainless-steel teeth. Called "workers" and "strippers," these drums perform the task of opening and fanning out the fibers to make a web spanning the width of the carder. A series of aprons reduces the gigantic wool sheet into ribbons that

are further attenuated into spaghetti-like strands that are wound onto giant rolls at the finished end of the carder. Called "pencil roving," the strands of wool resemble yarn but have no strength until twist is added at the spinning frame on the opposite side of the room.

The five-foot-wide spools of unspun roving are transferred to the Whitin spinning frame, built circa 1948. The frail strands of fiber are threaded through a pair of rollers and then down to the spinning ring. By making adjustments to a large and mysterious looking gearbox at the far end of the spinning frame, the operator, Patti or Ted, can determine the amount of draft (which sets the final diameter of the yarn) and twist (which sets the handle and drape of the yarn) the yarn will receive. The adjustments are subtle but can make or break a yarn run. When the entire line is in full-swing operation, the air rings with the mechanized whir of yarn winding onto bobbins. Airy, luminous hanks of yarn come off the skeiner at the end of the production line. Fondling them evokes a tactile reverie of the yarn's origin: sturdy, soft Border Leicester hogget ringlets and glossy mohair.

My Cormo wool travels a different route from sheep to skein. While Cormo makes a wonderful woolen yarn, having it spun worsted style produces a plush yarn with a sleek, honed surface, well suited to showing off stitch definition. Many years ago while attending SOAR (*Spin Off* magazine's autumn retreat), a hand-spinning retreat in held in BelAir, Michigan, I took a side trip to a fiber mill. Located on a rural road in the northern part of the state, Stonehedge Fiber Mill is the quintessential small-scale mill with big-time output. Founded by Deborah and Chuck McDermott in 1999, the mill originally operated with equipment salvaged from large-production textile mills. Continually growing and evolving, the mill now runs on a combination of refurbished vintage equipment along with new machines custom designed by Chuck himself to meet the needs of the operation.

Since my visit that year, each spring many oversize cartons of my flock's wool retrace my journey from Massachusetts to Michigan to Debbie, a fellow shepherd and a master yarn crafter. Her mill is a sprawling complex of buildings and sheds situated on a working farm. As a fellow fiber producer, Debbie understands well how much is at stake both emotionally and financially for yarn farms throughout the country and how her mill fills an important role for fiber farmers.

All good yarns begin with a conversation between the processor and the shepherd. Debbie and I often devise a game plan in the midst of lambing season over several phone calls. While juggling bottle lambs and grain deliveries, Deb and I discuss the clean yield of the wool clip and how to make the most of soft Cormo wool while managing its sometimes erratic springy tendencies.

Like Merino wool, Cormo wool often needs to be scoured twice to remove the lanolin before heading to the picker. In the scouring shed at Stonehedge, batches of wool suspended by nets of seine twine soak in a hot sudsy bath to remove grease and barnyard grit. In the next room washed wool is fed to the picker, which teases and fluffs the individual locks in preparation for carding. Puffy mounds of picked fiber are then ferried on a belt into the drum carder (a machine designed with help from the McDermotts), which further opens and aligns the wool to produce barrels of roving.

The fiber travels onward to the next building, where the pin-drafter grooms

multiple strands of roving through a series of eighty combs, to straighten and align the fibers. Pin-drafting, the last step before spinning, gives semiworsted yarns its smooth and polished finish. It also ensures consistent grist, the diameter of the finished yarn. An overhead creel feeds barrels of pin-drafted sliver to the spinning frame. Here the yarn at last begins to take shape.

"The machine will spin whatever it is given," Debbie explained. The more even the pin-drafted sliver, the more uniform the spinning. On the spinning frame a thick, less-drafted section of the sliver causes a hiccup. The twist is lazy and skips right over it, creating a slub in the spun yarn, the same way it does in hand spinning.

The ribbon of roving descends to a pair of rollers that form the drafting zone. Just as in hand spinning, it is here where the roving is attenuated to determine the yarn's final gauge. By adjusting gears, Debbie sets the amount of fiber receiving twist and also adjusts the spindle speed that regulates the amount of twist the fiber receives. As the machine purrs, bobbins fill with freshly spun singles. The plyer, a separate piece of equipment, marries pairs of singles to fill bobbins with buoyant two-ply yarn.

Observing the mechanical performance of the tasks I do by hand with hand cards, spinning wheel, and drop spindle is fascinating. As a hand spinner, my ability to test by hand the method of preparation, the amount of twist, and the ratio of fiber blend is a huge advantage in gauging the outcome when executing the same combination on a larger scale at the mill. While the processes are far from identical, the underlying elements of crafting yarn are the same whether performed by equipment or by hand. Fiber, grist, and twist.

THE ZEN OF DYEING

AFTER STICKY SUMMER DAYS, September is a fine time for dyeing wool. I throw open the doors of the dye studio, my work space in the garage attached to the studio. In clear weather, skeins dry quickly on the porch. The lambs grazing just outside the door keep me company. I talk nonsense to them while squishing bundles of yarn into the sink.

With fall yarn shares to ship and the fall fiber festivals on the horizon, wool in one form or another is never far from reach. I started Sheep Shares, a yarn and fiber CSA, in 2008. It's a model typically associated with farms that grow fruit and vegetables. Each season our farm's CSA members receive our produce—yarn and roving. Yarn-share members have the option of ordering their yarn shares either in the natural color of the sheep or custom dyed. For me it's exciting to dye yarn knowing exactly who's going to be knitting with it. Sometimes I know in advance what the project is, which adds to the fun. But dyeing yarn shares on top of dyeing yarn for fall fiber festivals is a tall order.

Practice has taught me to work without distraction in the dye studio. Although the process of dyeing yarn is simple, dyeing large lots of skeins in level

colors (nearly even solid hues) takes patience and focus. Get distracted for a few minutes at the wrong moment, and the whole process goes south. So, while dyeing, there are no phone calls, texts, or e-mails.

Prep work starts the night before. Since mixing dye powder into dye solution is my least favorite task and dyes are more safely handled in liquid form, I select my colors in advance. I boil water and stir dye powder into solutions. Plastic milk jugs, each holding a different color, line my workbench in preparation for dyeing day.

The colors for my fibers are inspired by what I see on the farm every day throughout the seasons: in the garden, in the fields, and in the sky. Cheerful squash blossoms, crimson sumac plumes, bronze sedge by the brook, a fence post freckled with blue-green lichen, the winter sky at dusk awash in hazy blue and smoky plum. They are germane to this place, my flock's home.

Once the colors are brewed, I soak the bundles of yarns in warm water. I

string skeins onto shoelace loops in bundles of ten or twelve. A nice warm bath removes the oils added to tame fibers in the spinning process and any residual barnyard grit. While the skeins soak, I scrub the kettles that will be my dye pots.

The next day starts with lighting a fire. Two large propane burners in the southeast corner of the room fire the dye pots. The grates on which the dye pots sit were once black but now wear a rusted patina. My kettles range in size, depending on the weight of the yarn and the size of the dye run. A forty-quart stockpot works well for runs of fingering weight and sock yarn. For sweater-sized skeins I use one-hundred-quart kettles from a beer-brewing supply house. The color-splattered wall behind the burners is a cheerful reminder of the hues of years gone by. I measure dye stock with a large graduated cylinder and pour it into the pots of water.

Kettle dyeing, otherwise known as immersion dyeing, is just what it sounds like: the yarn is immersed in a vat of water-and-dye solution. Acid, in the form of vinegar or citric acid crystals, and heat trigger the chemical process that bonds dye to fiber.

It is a simple process, but getting it right is sweaty, sometimes tedious work. Evenly dye-saturated solid color skeins take attention and muscle. The process is ruled by nonnegotiable chemical reactions. Protein fibers plus acid plus heat equal dye strike. The challenge is controlling the variables. For example, if the dye bath is cool but the yarn is still warmish from its bath, the heat (even just a little bit) makes the yarn grabby for color. And so you get splotches. Or, if the skeins sit too long in the same position in the dye pot as the heat rises, or if you don't stir the pot thoroughly enough before adding yarn, pockets of color become trapped here and there. And so you have streaks. If you walk away from a simmering dye pot to hang twenty skeins on the porch but then get sidetracked because the lambs are doing something funny, you return to a pot of yarn at a rolling boil. And so you have felt. It's best not to stray from the dye pots.

The looped skein bundles are strung through shoestrings. I suspend the shoestring leashes from hooks attached to steel cables directly above the pots. The cables are on a pulley and can be raised and lowered electronically using a hoist—the kind hunters use for hanging up a deer carcass. It took a rotator cuff injury for me to realize that while dry yarn is light and fluffy, wet skeins are heavy (though not quite as heavy as a dead deer).

Everything starts at room temperature—dye solution, water in the dye pot, yarn. Dye solution goes into the water. The yarn goes next. Light the burners. Every two minutes, lift and stir. I once toured a decommissioned commercial dye plant with my friend and fellow dyer Gail Callahan. Colleagues in color, we were itching to see what knowledge we could glean from this defunct plant that was about to be renovated for condos and artist lofts. From what we could tell, large dye houses use giant cylindrical tanks set into the floor and mechanical pumps to circulate the dye bath. Instead of pumps I have a strong upper right arm.

A dye bath starts off an opaque broth of color with only the tops of the submerged skeins visible. As the bath warms, the skeins steep. Lift and stir. Covering the pots speeds the rising stage of the bath. Ninety degrees. Lift and stir. One hundred twenty degrees. Lift and stir. The beer-brewing pots have thermometers on the front, which are helpful but not essential. I can tell what's happening by the sound of the water.

The pots hiss at 140 degrees. The skeins are too hot to handle. Wearing long rubber gloves, I use a metal slotted spoon to stir and maneuver skeins in the bath. Pulling all the skeins completely clear of the bath, I add the vinegar (or citric acid crystals), stir, and return the skeins to the pot. The colors begin to strike at 160 degrees. "Strike" is the word for when dyes chemically bond to fiber. I want color to strike all strands equally, everywhere, all at once. So I lift and stir continually now, first the pot on the left, then the pot on the right.

The hissing is louder when the bath hits the strike zone. Twenty degrees later, at 180, the strike is nearly complete, and the color is either level or not.

I once thought I could remedy an uneven dye strike by adding more dye, but it only darkened the dark spots more and created a blotchy mess. Many dye colors are formulated from tiny grains of different colored dye powder. Not all color components strike at the same temperature. Blues grab at lower temperatures, greens are fickle, and turquoise is temperamental. Red is downright stubborn. The dye bath needs every bit of an hour of the heat and sometimes another hit of acid to coax all color from the water.

As color molecules migrate from water to fiber, the once-opaque broth becomes translucent. When I can see the bottom, I can safely leave the pots to their own devices to prep the next round of skeins. A dye bath runs its course in sixty minutes from start to finish. After turning off the burners, I set the kettles full of skeins to cool on the concrete floor overnight.

In the morning I'm rewarded by a batch of color-saturated skeins in a completely exhausted dye bath. The yarn has absorbed all color from the dye bath—the water is now clear. I evaluate the results. Usually you can see where the shoestring loop has caused a slightly lighter spot where it compressed the yarn. And every skein has a dark and a light side where it rested against another skein in the pot during the dye process. I've learned to embrace the variations and imperfections that are the mark of skeins dyed by hand, not by a machine. Now I'm ready to fire up the pots again for another day's dyeing.

DYEING AND HAND SPINNING A SELF-STRIPING YARN

Hand spinners who dye their own roving have complete control over the outcome of color and pattern in the final project. I tell my students that learning to spin from roving you have dyed yourself opens all the doors of possibility for design. I used the following technique for dyeing and spinning a self-striping yarn for the Field and Sky Woven Tablet Cozy (see page 185). This process could easily be adapted to create a self-striping yarn for a scarf or a sweater.

DYEING THE FIBER

Dyeing long bands of color results in a self-striping yarn. You can spin one color without interruption for inches or feet or yards—depending on the diameter of the roving, the grist (thickness) of the yarn, and the dyed length of a color repeat. Long swaths of color maximize the reflective quality of longwools.

Before getting started, please read the Guidelines for Safely Dyeing Yarn and Fiber on page 305.

Materials

- 4 ounces pin-drafted roving: 60% Border Leicester, 40% mohair (You could substitute any longwool fiber for this process with good results.)
- Gaywool Dyes in four colors (Colors shown are Sugargum, Ivy, Meadow, Silver Birch.)
- Note: Since Gaywool dyes contain both dye and citric acid crystals combined, no additional citric acid or vinegar is needed for this process.

Equipment

- Eight one-quart mason jars
- A canning pot with rack and lid
- A heat source (such as a propane-fired camping stove)
- Chopsticks

Coil long portions of roving of equal amounts into eight quart-size mason jars as shown. The roving should be continuous, running from one jar to the next.

Add 2 cups of hot tap water to each jar.

Mix the dye solutions using separate jars or Pyrex measuring cups. For each color, add 1 teaspoon dye crystals to 1 cup boiling water for each color.

Set the mason jars into the canning rack. Decide on the color sequence for the striping. (For this project I repeated each color once.) Add ½ cup of each dye solution to each canning jar.

Place the rack with the canning jars into the pot. Add water to the canning pot just to the bottom of the steaming rack and cover the pot.

Using a heat source, bring the water to a simmer and maintain for 45 minutes. Allow the pot and jars to cool completely before handling the fiber.

Rinse the fiber in warm water and hang to dry.

Note: The dye solution will wick its way up the roving, but there will be areas of undyed roving where the roving travels from one jar to the next. If you want this segment of natural (undyed) roving

to be a design feature, you could intentionally leave longer loops of roving in between the colored jars.

SPINNING THE FIBER

In creating this yarn I wanted to maximize the qualities of longwool and mohair fibers that made them my first love as a hand spinner and dyer—luster and strength. Because longwools (for the sake of discussion I'm including mohair in this category even though technically wool is wool and mohair is mohair) are less crimpy than fine wools, yarn spun from them in a tightly twisted worsted style can feel wiry. I wanted to spin the fibers to maximize the luster and strength of both fibers while also minimizing the prickle factor.

Materials

- 4 ounces dyed pin-drafted roving: 60% Border Leicester, 40% mohair
- One spool of silk thread or mercerized cotton in a coordinating color for plying yarn

Equipment

- Spindle or spinning wheel

Specifications

- Yarn style: Single ply with a silk thread
- Method: Short forward draft, semi-worsted
- Wraps per inch: 15

The spinning goals were to keep the yarn soft, with good drape, yet also strong enough to handle some abrasion. Since a high twist can make longwools feel wiry and low-twist yarns break easily, I used a moderate amount of twist for spinning the single-ply yarn.

The second challenge was to create a self-striping warp. Since the color bands of the roving were long and uninterrupted, I wanted to maintain that look in the final yarn. Using a blue silk thread for the second ply balanced the twist while maintaining the effect of long color repeats in the yarn.

This project would be just as beautiful if you spun two singles and plied them together. The resulting colors would be more muted (since the dyed colors are now plied together), and the yarn would be heavier in gauge—which isn't a problem, since you will use it for the weft yarn in the following project. A heavier weft yarn will more quickly cover the warp.

TIPS FOR SPINNING

Dyed roving sometimes sticks together. Predraft the roving to ease the fibers apart to ensure smooth sailing.

Adjust the take up of your wheel to make sure the yarn receives sufficient twist. If spun too softly, the yarn will not be durable enough for this project. Periodically stop spinning and check the yarn by pulling a length from the bobbin. The yarn should feel sturdy and not come apart. Make sure you are not overtwisting the yarn.

Use the fingertips of your drafting hand to smooth the surface of the fibers as you draft them forward.

Once you've determined the grist and feel of your desired single, keep a sample strand of it taped or wound around an index card. Stop spinning periodically to refer to the sample for consistency. You can save the index card sample for reference for future projects.

FIELD AND SKY WOVEN TABLET COZY

Rugs are the most durable of woven textiles. I used basic rug weaving techniques to design this handwoven cozy for a standard-size iPad. This project can be tailored to fit any sized tablet or e-reader.

Finished Measurements

- Woven cloth: 20" × 10½" (50.5 cm × 26.5 cm)
- Woven cozy: 8" × 10½" (20 cm × 26.5 cm)

Materials

- 300 yards cotton carpet warp for the warp
- 4 ounces hand-spun wool-and-mohair yarn for the weft
- 8/2 unmercerized cotton yarn in a color analogous to the hand-spun yarn (for weaving the headers at the beginning and end)
- Thread in matching color and a needle
- A needle for the cotton carpet warp yarn
- A pair of well-worn blue jeans
- Note: Thrums and remnants left over from knitting and weaving projects are used for random accent stripes.

Equipment

- Four-harness floor or table loom (at least eighteen inches wide)
- One or more boat shuttle
- Sewing machine

Loom Specifications

- Width: 10½", in the reed
- Sett: 5 ends per inch (note: the warp ends are doubled—2 ends per shaft)
- Threading: straight draw (1, 2, 3, 4)

- Warp length: 3 yards (allows for loom waste and sampling)

SETTING UP THE LOOM

Wind a three-yard warp of 100 ends for the project. When dressing the loom, the first and last pair of warp ends are floating selvages (do not thread them through a heddle). The remaining 96 warp ends are threaded through the heddles and reed in pairs (in other words, thread 2 ends as 1 in each dent). Use a direct tie-up (one pedal for each harness) and a balanced 2:2 twill treadling pattern.

At the beginning and end of the project, weave a plain-weave 1½" header using the 8/2 cotton yarn. This is important for reducing the thickness of the cloth that you will machine sew when assembling the cozy.

Wind the hand-spun weft yarn onto bobbins. When weaving, throw each pick (pass of the shuttle) three times, beating on an open shed between each pick. Make sure to wrap the floating selvage as you exit each shed before entering the next shed. Treadle 2/2 balanced twill: 1, 2; 2, 3; 3, 4; 4, 1. The weft should entirely hide the warp threads.

If you weave the weft yarn as it was spun, the color sequence of your hand-dyed weft yarn will create stripes of alternating colors; I used scraps of yarn (one or two picks here and there) to add accent colors.

Weave a total of 20" (this includes the cotton headers at the beginning and end of the cloth).

SEWING THE COZY

Once you have cut the cloth from the loom, stabilize the ends by machine stitching two parallel rows from the cut raw edge of the cloth (one at least ¼" and another ½" from raw edge).

Cut a leg from a pair of blue jeans and then open the leg by cutting along the inseam. From the widest part of the leg, measure and cut a rectangle measuring 12" by 19". This is for the lining.

Turn under the two long sides of the denim lining by ¾" and press. The lining now measures 10½" wide.

Cut a rear pocket from the jeans leaving at least ½" for seam allowance for turning under raw edges.

Press under the raw edges of the pocket by ¼". See the diagram for pocket placement centered against the right side of the denim. Machine stitch along the edge to secure the pocket to the lining.

Cut the button from the waistband of the jeans, leaving about 1" of fabric on either side of the button. Determine the placement of the button by folding the woven cloth into thirds to form a pouch (as shown). Place the button on the front exterior of the pouch just below the edge of the folded flap. Fold under the cut edges of the button tab and hand sew the button in place. Note: Both the button tab and the buttonhole tab are thick, since they come from the waistband of a pair of jeans where many thicknesses of cloth are sewn together. You can easily reduce the thickness of these tabs by picking open the top stitching with a seam ripper. Remove unnecessary bits of fabric (from the waistband). Restitch the top stitching. It's easy and takes less than five minutes.

Cut the buttonhole tab from the jeans, leaving 1" to 2" of fabric extending from the buttonhole.

Place the lining and the woven fabric right sides together. Center the buttonhole tab between the lining and the woven fabric along the top edge, allowing the raw edge of the buttonhole tab to extend slightly past the raw edges of the fabrics. Machine stitch the lining to the woven fabric along the top edge, securing the buttonhole tab in place, and the bottom edge.

Turn the fabric right side out. The denim lining is now attached at the top and the bottom of the woven fabric. Using

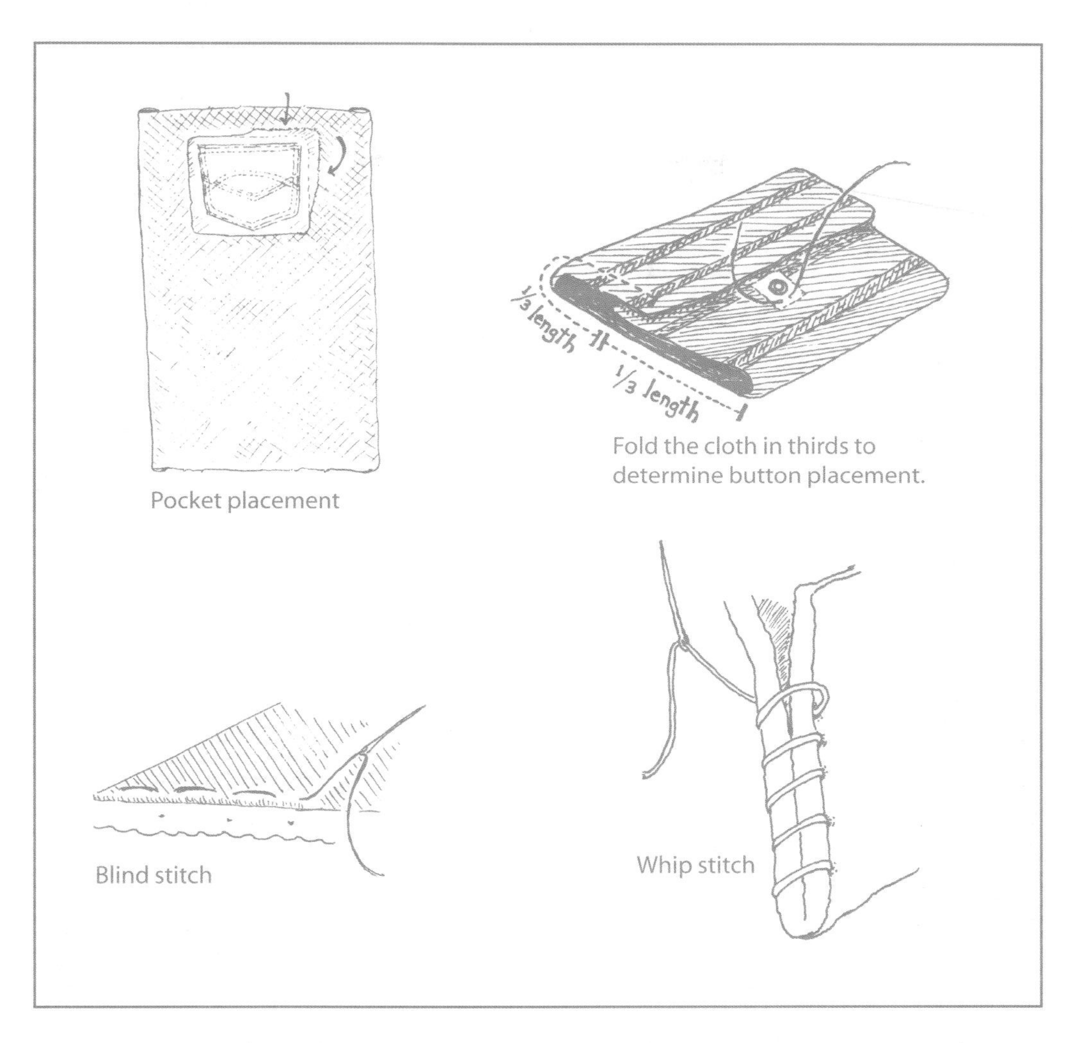

a blind stitch, hand sew the two sides of the lining to the woven cloth.

Fold the cozy in thirds so that it forms an envelope that will securely fit your tablet with the button fastened.

Using cotton carpet warp yarn from the warp, whip stitch the front of the pouch to the back of the pouch at the sides.

THE RAM IS HALF THE SWEATER

On a September morning I wake early and get up to close the bedroom window. Crisp, cold air means sweater season. I grab my favorite barn sweater from the chest and head downstairs.

The horizon is etched in rosy orange, but my view of the sloping pasture behind the house is murky. Although I can't see them, I hear the rams and wethers rustling about. The scuffling of many hooves on frosted turf makes a hollow, thudding sound. The hoof scuffling pauses. Then I hear the *thwack* of skulls colliding. There's a brief silence. The scuffling and thwacking resume, punctuated by random snorts and rumbling throaty snickers. *Scuffle, scuffle, scuffle. Thwack.* Our rams are testing each other's mettle, and the wethers are joining the fray.

Autumn's first bite triggers this behavior each year: the scrum for "top ram." It doesn't matter that the nearest ewe is half a mile away. The boys need no calendar to tell them breeding season is near. As the autumnal equinox approaches, receding daylight minutes and dropping temperatures create restless energy in the flock. The boys who have been amicable pasture mates all summer are suddenly extraornery and adversarial.

As daylight softly illuminated the pasture I saw this year's ram trifecta—Parsley, Teaberry, and Chai—is at the center of the scrimmage. Cinder, a black ram lamb, momentarily steps into the eye of the brawl but quickly reconsiders. His seniors hold a fifty-pound advantage. Cilantro, a robust, well-muscled, piebald wether, decides he, too, has something to prove. Our docile wethers scurry along with the group but keep to the periphery, steering clear of the blows. The group scrambles, snorts, and bashes all around the pasture while coffee brews.

Our largish flock of male sheep is atypical; most flock masters find a pair of rams ("a ram and a spare") sufficient. Ewes come into estrus every eighteen days in the fall. In theory, one ram would work the entire ewe flock for a period of thirty-six days in autumn. Then a second ram, the "cleanup ram," replaces the first for one last cycle to settle any ewes missed by ram number one. A fifty-four-day cycle would give at least three opportunities for ewes to be bred. At the end of the season, the boys take their leave. It's not unusual for a shepherd to keep a wether or two as ram companions.

Before experience taught us it would be wise to do otherwise, Mike and I

took a less conventional (and generally unadvised) approach to increasing our flock size. In our early years, we grew emotionally attached to all of the lambs—both rams and ewes—born on the farm. Neutering ram lambs to keep with our flock sires seemed like a good idea at the time. They produce jumbo-sized, spectacular fleeces. But they have jumbo-sized appetites to match and are a handful. As summer wanes, testosterone-driven challenges rise among all the boys.

For this reason, our boys spend the summer in the northwestern reaches of the farm, rotating from the pasture just outside our door, the fields along the driveway, to the birch tree lot and then back again. It's a two-pronged strategy. The boys munch and maintain the land farthest from the farm proper while keeping a respectable distance from the ewes, who spend their summers grazing the fields downhill (and downwind).

I can trace my flock's history through our succession of rams. Ram selection is one of the most important decisions a shepherd makes in shaping a flock. Shepherds say "the ram is half the flock" because a ram is the cornerstone of a flock. While a ewe passes her traits only to her own offspring, the ram passes

his characteristics on to every lamb he sires. An awesome ram is the fastest way to improve a flock. Since wool is a highly heritable trait, a solid ram is the fastest way to improve the quality of a wool clip.

Rams can also be the most temperamental members of the flock. The word *ram* can mean "striking with great force" for good reason. I filed cautionary tales of broken gates and bruised hips into the back of my mind as a novice shepherd seeking our first herd sire in 2000. My favorite sheep reference book, *Raising Sheep the Modern Way*, shows its author, the famed shepherd Paula Simmons, being chased by her favorite ram. Despite my growing skills in working with my small ewe flock, when it came to keeping a ram, I had low sheep confidence.

Fortunately, before I became a ramaphobic wreck, my shearer, Andy, pointed me in the direction of Isaac, a seasoned Border Leicester ram nearing retirement age at a nearby farm. Isaac had a reliable track record and mild manners. Although his fleece was rather open and wavy and was coarsening with age, he was sturdy. What Isaac lacked in ideal wool characteristics, he more than made up for with a mellow temperament, which made him the perfect starter ram for me.

Isaac's gentle demeanor quickly assuaged my apprehension. In our early years at Springdelle Farm, our pastures filled with Isaac's lambs—more rams than ewes, to my chagrin. One of the pitfalls of sheep breeding is that within a couple of years, ewes sired by the flock ram come of breeding age. The only way out of the genetic cul-de-sac is to widen the gene pool. To avoid line breeding Isaac to his own daughters, we were soon in search of another ram.

By this time my goal had sharpened: to improve my flock's wool clip. The national Border Leicester show was being held that year in Springfield. Breeders nationwide bring their most promising rams, ewes, and lambs to the show ring. Prospective buyers see the best each flock has to offer and can compare flocks of the same breed side by side. Sheep shows attract many casual visitors, so it's easy to browse the show pens without telegraphing that you are seriously in the market for a sheep until you've narrowed down the prospects. I was looking for a white Border Leicester replacement ram lamb. I had come to the right place.

A ram lamb from a farm in Groton, New York, stood out after two surveys of the pens. This youngster had good lines and an impressive robe of tight, glossy ringlets. He had presence. There was something composed in the way he studied me as I studied him. I sought out the breeder, who mentioned that he was slated to enter the ring the following day.

I looked more closely. His mouth was sound, with a proper bite (neither under- nor overshot). His legs were straight, and he stood high and strong on his pasterns. While the breeder held the halter, I closely inspected each hoof and queried the breeder about the flock health. Was there a history of foot rot? What was the flock's scrapie status?

After kicking the tires, it was time to look under the hood. When shopping for a show-quality ram, one must inspect the equipment. On my knees in the straw of the holding pen, with countless people milling around, I performed an essential inspection that any sane breeder should perform (but not something I could ever have imagined myself doing five years earlier, much less with an audience). I had read up on this very topic in advance, so I'd know what to look for. Both testes present and fully descended? Check. Testes equal in size? (Each about the size of a walnut, in case you're wondering.) Check.

Sheep industry literature advises that the scrotal circumference of a ram lamb is a key indicator of future performance. I checked the necessary dimensions, using a piece of cotton twine from my pocket. Thirty centimeters, a respectable size for a junior ram lamb. The ram was remarkably obliging. We had gained onlookers.

On his pedigree I recognized the name of a well-respected Canadian flock. Given the recently tightened restrictions on sheep crossing the border (for biological security) the Canadian lineage was another plus.

I liked everything about this ram lamb, so I negotiated a sale price with the owner. The next day I called her to see how he fared in the ring. With ribbons under his halter—first place in the ram-lamb division and an award in the fleece competition—I was glad I had closed the deal on the sale price the day before. Opi made his way to Springdelle in 2003, following in Isaac's hoofsteps as our Border Leicester herd sire.

That very same year we decided to add Cormo sheep to the flock, starting the entire process of ram (and ewe) selection afresh. While many of the same considerations for choosing a ram applied, this time around the fiber characteristics I was looking for were entirely different. I was more focused on raising fine wools of high quality and yield. By the time our first Cormo males joined the flock—Trumpet, a ram lamb, and Jack, a friendly wether for his sidekick—it was with far less trepidation on my part.

My confidence in sheep management has grown, which is a very good thing, and over the years, we've added more rams that have made their stamp on the farm's wool. Nonetheless, every fall it comes time to plan for the future spring, pairing ewes and rams for breeding.

When planning woolly matchmaking, compatibility criteria start with keeping close kin separate. Trumpet's daughters are suitable pairings with Teaberry. Leicester ewes related to Opi might be matched to either Cormo ram, but I ponder the outcome of a Cormo-Leicester cross. Parsley is an eligible suitor for Cocoa and any of Cocoa's daughters not sired by Trumpet. To make things more interesting, we eventually recruit a fine-wool Moorit ram, Chai, to the flock. Brown is a recessive color trait. Wouldn't it be fun if a Cocoa daughter threw a Moorit lamb? It quickly gets complicated keeping family trees straight.

I rely on my flock record book as an almanac. Each ewe has her very own page, noting lineage, birth number (single, twin, triplet, or quadruplet), and other qualities. Good or bad, everything gets noted here: Was she a good mother? Did last year's lambs thrive? Is her udder free from defect? I pay homage to each of my original ewes by keeping the best of her daughters.

If there is more than one potential pairing for a ewe, I consult the fiber archive that holds the history of my flock as written in wool. This is one reason why I save a single lock of wool from *every* sheep on *every* shearing day. Organized chronologically, each box represents the wool vintage in any given year.

Sorting through the unwashed locks of wool spread out on the table before me gives me the big picture. It's perhaps the best illustration of how the ram factor shapes the story of the wool. There's a marked change in pearlescent sheen and curliness once Opi's offspring reach shearing age. The hogget fleece (a sheep's yearling first fleece) samples of Trumpet's first sons and daughters reflect his influence—superfine, with crimp so tiny you need a magnifying glass to count it. Teaberry's daughters' hogget fleeces show more length and a bolder crimp. When it comes to wool, the apple doesn't fall far from the tree.

A separate binder holds my notes from shearing days, where we've kept track of the most promising fleeces as they come off the sheep. The grease weight of a roughly skirted fleece is a good way to compare yield. The Trumpet-Parsley line throws denser fleeces with a shorter staple. Teaberry's family fleeces have a slightly more open characteristic but also more length.

Ultimately, my hope is to pick the ewes that will deliver with no fuss and be good mothers to healthy lambs who will carry the most desirable yarn-defining characteristics of both parents. That's the goal. An autumn afternoon is well spent with a cup of tea, reviewing notes and locks from years gone by, planning for my flock's future.

MULE
3010

GOATS ARE BORN LOOKING FOR TROUBLE

I CONSULTED ANDY again when I wanted to explore the possibility of adding Angora goats to my flock of sheep. With a chuckle he advised, "goats are born looking for somewhere to get into trouble; sheep are born looking for a place to die." This didn't stop me from testing the waters and acquiring a pair of friendly, wethered buck goats. Butch and Sundance came to me as yearlings from Bob Ramirez's flock at Keldaby Farm in nearby Colrain, Massachusetts. Rather than dive in headfirst with a flock of does for breeding, I thought it best to test the water with these good-natured wethers, who were inquisitive and easy to handle. The handling part is important with these animals.

Next to alpaca and silk, mohair is my favorite fiber for blending with wool. It seemed logical to have a go at raising this fiber myself. I quickly learned that this is easier said than done.

Although goats and sheep can be raised side by side, the management of the animals is different. Angora goats are mohair-growing machines. Because of their high fleece output, they are sheared twice a year. Each shearing nets a fleece that

could be up to a quarter of the goat's total body weight and four to six inches in length. The flip side to Angora goats' high fiber output is the tremendous metabolic demand it places on the animal. A goat puts all of its energy into generating fleece; in fact, a goat will suffer before compromising its fiber. The breed originates from Turkey, and raising them in cold, wet New England takes special consideration. Less hardy than their sheep counterparts, they need more protection from wind and wet, especially when recently sheared.

In addition to their fleece's needing managing, Angora goats need frequent hoof trims. At Bob's advice, I bought a metal trimming stand. The boys were used to jumping up on the table and being fitted into a stanchion to hold them in place while I worked on their feet. While I had them stanchioned, I trimmed around their eyes, so they could see; I trimmed their beards to keep them tidy and trimmed the mohair from around their pizzles to prevent pizzle rot. Bob had also counseled that Angora goats can be lice magnets. Treating them for

external parasites is just as important as treating them for internal parasites. Their four-week maintenance program included a thorough combing and parting of their wool for a close examination for brown goat lice and spraying their armpits if needed with a water-based insecticide, permethrin. The chore of lice management nearly put me off the idea of goats altogether until Bob explained that lice are species specific. Goat lice cannot live on sheep—or on humans. That didn't stop me from scratching my head while working on the boys. Also, external parasites are easily managed, especially in small flocks.

The need to handle the boys often taught me a lot about the differences between sheep and goat psychology. As close as I was to my sheep, they still perceived me as "other," and their flight instinct would always kick in by default whenever I entered the pen. They couldn't help it. The goats seemed to perceive me as equal. If I was working in the barn, they were more interested in whatever I was doing than in what the sheep were doing. Rather than retreat, they instinctively sought ways to interact. If I was wearing a wool cap and bent over to fill a bucket, Butch would pull it off my head and shake it. If I drove the Mule, our all-terrain buggy for getting around the farm, out into the pasture to fill the mineral feeders, the boys would hop into the cab to ride shotgun with me, or scramble up into the bed in the back, to inspect my cargo.

Their agility and love of the precarious led to all sorts of interesting shenanigans. The boys easily scaled the concrete feed bunk in the open barn and would tightrope walk along the top of it. From there, they could sail over the barn partitions that separated one pen of sheep from another. It made containing them difficult. If they wanted to be somewhere, they got there. They figured out how to escape the net fence enclosures by listening to the snap made by the pulse of the electric fence. In the second between pulses, they could flip up the fence at the ground and pop right underneath it.

They seemed to have a dim view of the sheep as unworthy companions. In the barn hierarchy, the goats let it be known that they held the upper hand. In winter, when everyone is fed in close quarters at a series of long feeders, the two goats would commandeer sixteen feet of feeder space, using their pointed headsets to fend off interlopers. At times the goats were so mean we simply had to place them in their own building, which drove them nuts. As much as they disdained sheep, they couldn't stand not being able to see what the sheep were doing. They had to be at the center of the action.

I'm certain anyone who has ever kept goats would agree—although their antics are sometimes maddening, they are mostly endearing. Butch and Sundance were gregarious participants in barn chores, never missing a chance to pitch in.

On several occasions they've fed the entire flock for us by popping open the gates that kept the sheep from the hay. Thanks to Butch and Sundance, one evening the flock had free run of the barn—we arrived to find pulled apart hay bales, poop, and baling twine strewn everywhere.

On another day, they released the entire flock out into the neighborhood. The sheep kept close to the farm, but the boys wandered up the road where I found them "weeding" our neighbor's perennial garden.

One day I came to the barn to find Butch who, in tidying up the barn, had hooked himself like a fish through the side of his mouth on a stray bungee cord.

While pruning the multiflora rose along the lane one afternoon, Sunny snagged his ringlets in the thorny tentacles. Extracting him one branch at a time was a prickly project. He was patient and grateful.

In summers when I delivered water to them in the upper pasture, they would ambush the Mule; while I filled the stock tank, they clambered into the driver's seat, ransacking the glove box and pulling the leather key fob from the ignition.

Sadly for both of the bucks, they each eventually got into serious trouble that led to their demises. For Sundance it was a tragic tangle one night in the electric fence. Butch was horribly lonely without a goat companion, so we adopted Gypsy, a little doe, to keep him company. Then two years later, Butch dislocated his hip while apple picking at the tip-top of a stone wall. The vet was unable to fix him, and so we had to put him down.

We still have Gypsy, and though she misses Butch, she seems more at ease with (and less bothersome toward) the sheep than the boys had been. Like the boys, she is small, agile, and able to climb and escape from almost any pen in the barn. From time to time we consider adopting another goat friend for her but have decided, at least for the moment, that one goat is enough.

DYEING, COMBING, AND SPINNING MOHAIR

Unlike wool, mohair quality is graded according to the age of the goat from which it is sheared. A kid fleece is from a baby goat. It is the finest, silkiest fleece the goat will ever produce. Yearling mohair comes from the second and third shearings, done while the goat is still young. It is also highly desirable for its softness and luster and is also longer than the first shearing. As the goat matures, its fleece coarsens and dulls. Adult mohair is still a useful, durable fiber, especially as upholstery and rug yarn.

Like wool fibers, individual mohair fibers are covered by microscopic scales. The scale surface is longer and smoother than on wool. The scales reflect light, giving mohair its characteristic sheen. While mohair locks are curly, they are not springy. Yarns spun from them are less elastic than wool yarns. When combined with wool, mohair adds both strength and shine.

Following are notes on dyeing, combing, and spinning mohair fibers. This is the process I followed for creating the yarn used for the Sundance Scarf on page 215.

DYEING MOHAIR LOCKS

The dye process that I used to make the mohair yarn used in the Sundance Scarf is called low-water immersion, and it is a simple way to create variegated colors for locks, roving, or yarn.

Before getting started, please read and follow the Guidelines for Safely Dyeing Yarn and Fiber on page 305.

Materials

- 4 ounces clean mohair locks
- PRO WashFast Acid Dye: Tiger Lily, Mahogany, Chile Pepper
- ¼ cup white vinegar

Equipment

- Pyrex vessel
- Colander or nylon mesh laundry bag
- Shallow stainless or enamel pot or pan (dedicated for craft purposes)
- Candy thermometer
- Tongs, chopsticks, or spoon for manipulating fiber
- Outdoor propane burner or indoor cooktop

Presoak the mohair locks in a basin of warm water with ¼ cup vinegar. The vinegar is necessary for the protein fibers of the mohair to bond with the dye molecules in the pot. Soak the locks for at least one hour.

Mix the dye solutions. Measure 1¾ teaspoons of dye powder into a Pyrex vessel. Dissolve the dye powder with 2 cups of boiling water. Stir thoroughly.

Remove the mohair from the presoak and drain using a colander or mesh laundry bag.

Spread the mohair locks evenly in a shallow enamel pan. Add just enough water to half submerge the locks. Bring the water to a simmer over a low heat.

At striking temperature (160 degrees), pour small amounts of different colors of dye solution randomly, one at a time, directly onto the fiber. Pour small amounts of each color onto a different area. Start with your lightest colors and work toward deeper shades; use the colors sparingly.

Wait for each color to strike before applying the next color. You can tell when the fibers have grabbed all the dye when the water in the bath is clear.

Once you've applied the dyes, allow the fiber to simmer for forty-five minutes, keeping the water temperature no higher than 180 degrees. This takes watching, as the water level is low.

Do not stir the fibers and do not let the water come to a rolling boil. This would felt the locks together.

Allow the fibers to cool in the dye pot. Rinse the dyed locks using a mesh laundry bag, taking care not to agitate them. Spread the dyed mohair on a rack or nylon screen to air dry.

COMBING MOHAIR

Combing is the method of choice for preparing mohair fibers for spinning. A combed preparation maximizes the fiber's intrinsic qualities: luster and drape. Combing with wool combs aligns the slippery fibers for spinning a smooth, shiny yarn.

Wool combs are lethal-looking tools. They consist of a wooden block, called the head, attached to a handle. Long, sharp steel teeth, called tines, line the head of the comb. The "pitch" of a wool comb refers to its number of rows of tines. The more rows of tines, the higher the pitch and the finer the fiber the combs will handle. I used a pair of 2-pitch St. Blaise wool combs to prepare Butch and Sundance's locks for spinning.

Combs are used in pairs. One comb is fastened to a table with a clamp, the other is held in your hand. Clean locks of mohair (or wool) are loaded onto the clamped comb one lock at a time. To do so, hold a lock by the tip and flick it downward onto the tines with the comb in your hand, catching the butt end of each lock on the tines of the comb

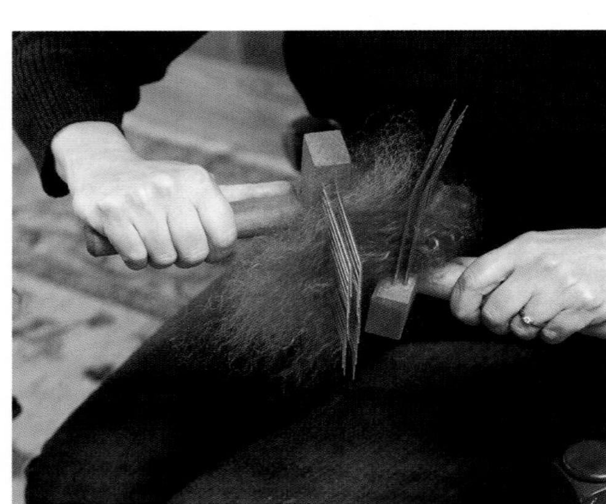

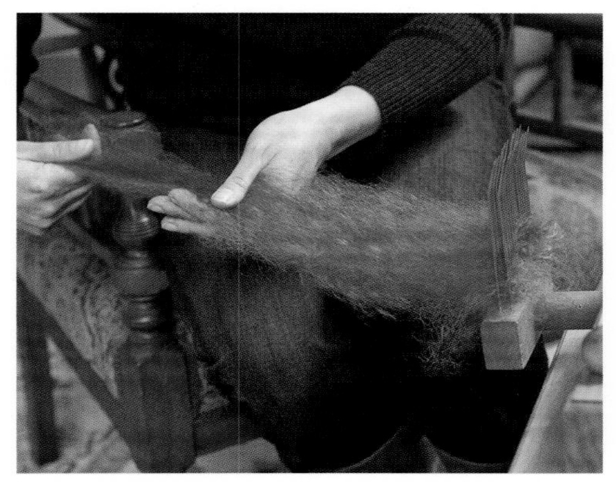
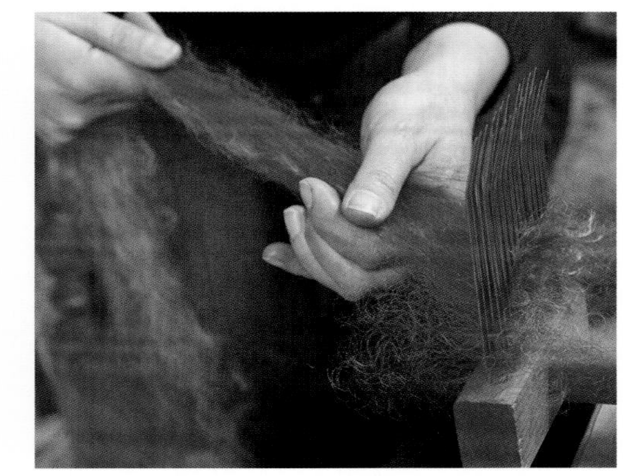

secured to the table. This step is called loading the comb.

The combing process is a patient series of passes. With a sweeping motion the "passing" comb comes down and combs through the tips of the mohair locks on the first pass, just opening them up. The second pass goes a bit deeper into the locks, further opening them. In a series of sweeping motions, you comb the locks, each pass going further into the staple with slightly more force in each pass—they way you comb your hair first thing in the morning.

Eventually, very little is left on the stationary comb. The remaining dross (noils, short pieces, bits of seed) can be discarded. The two combs exchange places and the combing process is repeated.

With each series of comb passes, the fibers get smoother, more aligned, and more uniform. It's a winnowing process, leaving the longest fibers on the combs, all similar in length. From the stationary comb, the slick, smooth fibers can be drawn out to a uniform thickness, either by eye or by using a tool called a diz. Traditionally made from a longitudinal slice of cow horn (though mine is made of wood), a diz is a concave disc with a hole drilled through. The fibers are pulled through the hole to create a long sliver of combed top of even diameter ready for worsted-style spinning.

A few helpful tips: First, the fiber should be well scoured and dry before combing. Lanolin makes the fibers sticky and will also stick to the tines of your combs. If you want to add color, dye the locks before combing. Combing creates static electricity. Use a spray bottle to lightly mist the locks with water before combing. A spritz of olive oil (I use a shot of spray PAM olive oil) conditions the fibers, making them glide easily. Take your time. This is not a task for when you're tired or distracted. If you catch the tip of your finger on the teeth of the wool combs, you will draw blood. If you are the impatient type, this may not be the process for you. Combing is meditative, much like hand carding fleece. Take it slowly, one lock at a time. The secret to worsted style yarn is the preparation.

SPINNING WORSTED YARN

Most hand spinners are taught worsted-style spinning and don't even realize it. Worsted spinning produces a smooth, firm yarn with little loft. Whether done on a drop spindle or on a spinning wheel or an industrial spinning frame, the fibers are parallel as they receive twist.

Butch and Sundance's combed locks spun obligingly into a sleek yarn for this project. Divide the sliver into two equal amounts—one for each ply. The medium-size whorl on my Lendrum wheel provided just the right amount of twist (the smaller whorl energized the yarn too much; the larger whorl didn't provide enough twist—the yarn would have slipped apart). Because combed mohair is slippery, no predrafting is needed. I used a light to moderate tension on my bobbin (just enough to wind on the yarn without pulling the yarn out of my hands).

When spinning worsted style, it's all about managing the twist. Keep your fiber supply in order; the sliver should set in your hands and then run down to one side (a basket on the floor beside you works well). The hand closest to your body gently grasps the sliver. The hand closest to the spinning wheel orifice drafts the fiber forward and controls the twist. A pinch of the thumb and forefinger serves as gatekeeper of the drafting zone—the fan-shaped triangle of fiber between your two hands. The twist may not enter the drafting zone until you say so.

The forward hand pinches the unspun fiber and draws a small amount forward toward the wheel. Then it slides back, while never releasing contact with the fibers. Those fingers are the gatekeepers of twist; the twist must never enter the drafting triangle. The fiber-supply hand feeds a steady fresh supply of sliver into the drafting triangle. It must not grasp the sliver too firmly or it will disrupt the arrangement of the fibers. Both hands move continually, simultaneously. The feet treadle. Without overthinking it, find a rhythm, a certain number of treadles per drafting motion. The smoother your draft, the more consistent the yarn. Twisted yarn, smooth and shiny, runs into the orifice and winds onto the bobbin.

Spin the sliver on two bobbins. Using a lazy kate, ply the two singles together taking care not to overply. Wind the plied yarn into a skein and give the skein a warm soak to set the twist. I like to add a dab of hair conditioner to the soak. Gently press the excess water from the skein and hang to dry overnight.

SUNDANCE SCARF

Designed by Marnie MacLean

Marnie's lace scarf is shown in two versions. The mill-spun version is made with my Cormo Alpaca Lace. The high percentage of alpaca (40%) counterbalances the springy Cormo wool. This gives the scarf nice drape, keeps the lace design open and well defined, but still has the Cormo memory, so it won't stretch out of shape with wear.

The hand-dyed, hand-spun version is named for Sundance, my sweet but sometimes naughty Angora buck, and made from the yearling clips of Sundance and his brother and partner in crime, Butch. Unlike the mill-spun version, the mohair version has luster and a distinct halo.

The carefree style of this lace scarf makes it a perfect accent with a sweater or jean jacket. Using only one skein, the scarf is lightweight yet warm.

Like most lace, this scarf needs to be fully blocked upon completion. I recommend gently washing the finished scarf and pinning it out while it is still wet.

Finished Measurements

- Approximately 48" (122 cm) long by 8" (20 cm) wide.

Yarn

- Version 1: Foxfire Fiber & Designs, Cormo Alpaca Lace (60% Cormo wool, 40% alpaca; 240 yd [220 m]): 1 skein. Shown in Winterberry.
- Version 2: Hand-spun mohair (single ply, 22 wpi), approximately 240 yards

Needles

- One set size US 6 (4 mm) straight needles

Notions

- Stitch markers (if desired)
- Tapestry needle
- Blocking supplies

Gauge

14½ sts and 26 rows = 4" (10 cm) in lace pattern. Adjust needle size as necessary to obtain correct gauge.

Note: Gauge is not terribly important for scarves, but different gauges may effect yardage requirements.

Modifying Finished Size

Add stitches in multiples of 24 to widen the piece. Work extra repeats of charts B and/or D to lengthen the scarf. If you change the size of the scarf, the yardage needed will also have to be adjusted.

Reading the Charts

Charts A and E show RS and WS rows. RS rows are read right to left. WS rows are read left to right.

Charts B, C, and D show RS rows

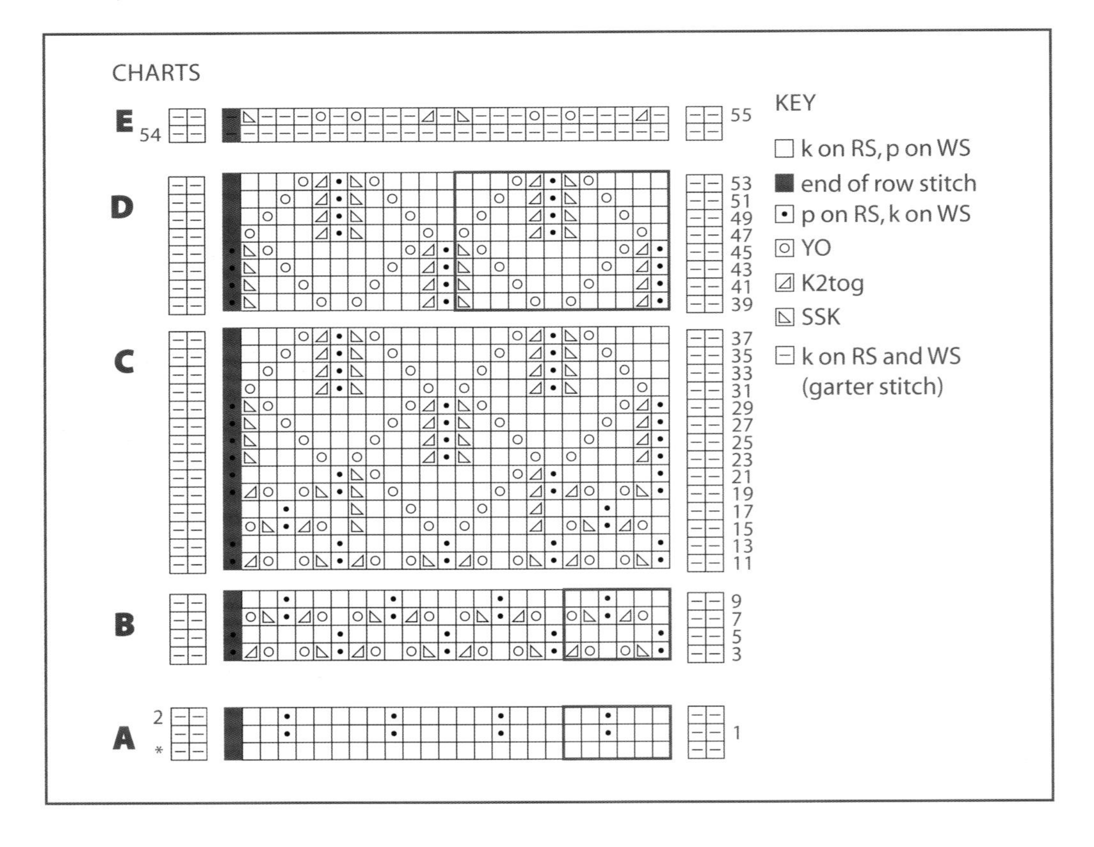

only. All WS rows should be worked as follows: K2, knit all knit sts, purl all other sts to last 2 sts, k2. Note that a WS row needs to be added between all charts *except* between charts D and E.

When reading RS rows, the first and last two stitches are always knit. If the pattern repeats, there will be a red line around one complete repeat. The third-to-last stitch is red and is the additional stitch worked after completing all full repeats across.

I prefer to place stitch markers between repeats, but this is optional according to preference.

SCARF

Working a provisional cast on, CO 29 sts.

Work chart A once, chart B nine times, then charts C through E once. BO then remove the provisional cast on to pick up sts. Work the second side in the same order as the first: Work chart A once (do not work the foundation row on the second side), chart B nine times, then charts C through E once and BO.

Hint: Wool yarns can often be wet felted. When you join to start in the other direction, you can felt your tail from the CO to your working yarn to avoid having additional ends to weave in.

CHART A

Foundation Row (WS): K2, purl until 2 sts rem, k2.

Row 1 [RS]: K2, *k3, p1, k2, repeat from * until 3 sts rem, k3.

Row 2: K2, work in patt until 2 sts rem, k2.

CHART B

Row 3 (RS): K2, *p1, ssk, yo, k1, yo, k2tog, repeat from * until 3 sts rem, p1, k2.

Row 4 and all WS rows: K2, knit all knit sts, purl all other sts until 2 sts rem, k2.

Row 5: K2, *p1, k5, repeat from * until 3 sts rem, p1, k2.

Row 7: K2, *k1, yo, k2tog, p1, ssk, yo, repeat from * until 3 sts rem, k3.

Row 9: K2, *k3, p1, k2, repeat from * until 3 sts rem, k3.

Row 10: K2, knit all knit sts, purl all other sts until 2 sts rem, k2.

Repeat rows 3–10 eight more times for a total of 9 repeats of chart B.

CHART C

Row 11: K2, *p1, ssk, yo, k1, yo, k2tog, repeat from *, p1, k2.

Row 12 and all WS rows: K2, knit all knit sts, purl all other sts until 2 sts rem, k2.

Row 13: K2, p1, *k5, p1, repeat from * until 2 sts rem, k2.

Row 15: K3, yo, k2tog, p1, ssk, yo, k1, k2tog, k3, yo, k1, yo, k3, ssk, k1, yo, k2tog, p1, ssk, yo, k3.
Row 17: K5, p1, k3, k2tog, k2, yo, k3, yo, k2, ssk, k3, p1, k5.
Row 19: K2, p1, ssk, yo, k1, yo, k2tog, p1, k2tog, k1, yo, k5, yo, k1, ssk, p1, ssk, yo, k1, yo, k2tog, p1, k2.
Row 21: K2, p1, k5, p1, k2tog, yo, k7, yo, ssk, p1, k5, p1, k2.
Row 23: K2, p1, k2tog, k3, yo, k1, yo, k3, ssk, p1, k2tog, k3, yo, k1, yo, k3, ssk, p1, k2.
Row 25: K2, p1, k2tog, k2, yo, k3, yo, k2, ssk, p1, k2tog, k2, yo, k3, yo, k2, ssk, p1, k2.
Row 27: K2, p1, k2tog, k1, yo, k5, yo, k1, ssk, p1, k2tog, k1, yo, k5, yo, k1, ssk, p1, k2.
Row 29: K2, p1, k2tog, yo, k7, yo, ssk, p1, k2tog, yo, k7, yo, ssk, p1, k2.
Row 31: K2, *k1, yo, k3, ssk, p1, k2tog, k3, yo, repeat from *, k3.
Row 33: K2, *k2, yo, k2, ssk, p1, k2tog, k2, yo, k1, repeat from *, k3.
Row 35: K2, *k3, yo, k1, ssk, p1, k2tog, k1, yo, k2, repeat from *, k3.
Row 37: K2, *k4, yo, ssk, p1, k2tog, yo, k3, repeat from *, k3.

CHART D

Row 39: K2, *p1, k2tog, k3, yo, k1, yo, k3, ssk, repeat from *, p1, k2.
Row 40 and all WS rows: K2, knit all knit sts, purl all other sts until 2 sts rem, k2.
Row 41: K2, *p1, k2tog, k2, yo, k3, yo, k2, ssk, repeat from *, p1, k2.
Row 43: K2, *p1, k2tog, k1, yo, k5, yo, k1, ssk, repeat from *, p1, k2.
Row 45: K2, *p1, k2tog, yo, k7, yo, ssk, repeat from *, p1, k2.
Row 47: K2, *k1, yo, k3, ssk, p1, k2tog, k3, yo, repeat from *, k3.
Row 49: K2, *k2, yo, k2, ssk, p1, k2tog, k2, yo, k1, repeat from *, k3.
Row 51: K2, *k3, yo, k1, ssk, p1, k2tog, k1, yo, k2, repeat from *, k3.
Row 53: K2, *k4, yo, ssk, p1, k2tog, yo, k3, repeat from *, k3.

CHART E

Row 54 (WS): Knit.
Row 55: K2, *k1, k2tog, k3, yo, k1, yo, k3, ssk, repeat from *, k3.

Repeat rows 54–55 two times. BO loosely, knitwise. Cut yarn.

FINISHING

Weave in tails and block, being sure to allow the ends of the scarf to maintain their natural scalloped shape.

FLEECE
SALES

FOXFIRE

FOXFIRE
FOXFIRE

NEW YORK STATE SHEEP AND WOOL FESTIVAL

ONCE ALL THE SKEINS have been dyed, it's time to send them out into the world. One of the biggest venues for showcasing my yarn is the New York State Sheep and Wool Festival.

On the third weekend of October each year, the Dutchess County fairground in New York's Hudson River valley town of Rhinebeck is a mecca for fiberists and yarnistas from the Northeast and beyond. Among the fall fiber festivals of the region, this is the biggest, drawing vendors from around the country and visitors from around the world. Serious fiber aficionados book hotel rooms a year in advance. For knitting groups and spinning guilds, this is a must-attend event.

My Rhinebeck initiation was ten years ago—at first as a "civilian" joined by my sisters Trish and Kathleen. We share a mutual passion for fiber in all forms.

My experience of attending the New York State Sheep and Wool Festival morphed from spectator to vendor once I had firmly established my flock and yarn business. Although my focus and objectives have changed, it is still one of my favorite weekends. The economic viability of my farm, and of many of

my fiber-farming peers, rides on this festival. As the fiber festival season winds down, a good weekend of sales here is crucial for ending the season solidly in the black.

We arrive the day before the festival to set up our booth. Mike has refined the art of packing the van for this show. The trick is to load the items needed first upon arrival—booth shelves, tables, lighting—on top. Heavy bins of yarn are the base of the load. Bags of roving and fleece get stuffed into the spaces between items. In case of a collision, they would handily serve as airbags for my booth mannequins.

When attending a fiber festival as a spectator, you pack as little as possible, leaving plenty of room for "the haul" on the return trip: fleeces, antique spinning wheels, masonry sheep statues, roving, yarn, and for us one year, an Angora bunny. The equation is exactly reversed if you are a vendor. Then you are trying to fill every square inch of the rental van with product while leaving a slice of visibility through the rearview mirror. At the weekend's end, a van filled with empty bins spells success.

Upon arrival, vendors driving rental trucks and towing trailers jockey for parking spaces nearest to the doors of the various exhibit halls. It's chaotic and unruly. Vendor etiquette says unload quickly and then move your van so others can do the same. Using hand trucks and dollies, vendors haul their booth gear and boxes of wool to their assigned show space. Arrival time is a convivial reunion. Some of the vendors near our booth have been "neighbors" at this festival for years. This may be the only time we see each other, so there are lots of hugs and hellos. The Testa family of Weston Farm can be counted on for lively tunes during setup. Nancy, of Long Ridge Farm, waves a big hello as she passes. On the way to the restroom, I bump into Linda Cortright, editor of *Wild Fibers* magazine. Jenn from Spirit Trail Fiberworks is just around the corner, and Alice from Fox Hill Farm is a block away. Spirits are high as we ferry and unpack for the entire afternoon. On Saturday morning we come back early and put the finishing touches on the booth.

The festival opens at nine o'clock Saturday morning. Veteran fiberists know an early start avoids the traffic snarl on Route 9 that will get backed up to the Kingston bridge by midmorning. So there are lines at all three gates before the show starts.

For first-time visitors, this show can be daunting. The fairground covers more than 160 acres, but fiber purveyors, food concessions, and livestock exhibits are concentrated in a warren of cattle barns, pavilions, trailers, and pop-up tents in the center of the fairground.

A plan of attack depends on your interests. Casual visitors may stroll the buildings in order, taking in the sheep breed informational barns and then perusing the fiber wares.

But the hard-core knitters and fiberists come armed with lists and a game plan. In order to get first pick of the swag they make favorite vendors the first stops. They wear backpacks and carry oversized tote bags and clutch shopping lists. Knitters have a mental checklist of yardages and gauge.

The vendor halls quickly become a buzzing hive of activity. Fiber farms, spindle makers, indie dyers, rug hookers, weavers, and commercial yarn purveyors are side by side on both sides of every aisle. It's a mélange of color, texture, and smell. You may catch the musk of ram's fleece in a fiber-farm booth or

US

a whiff of vinegar emanating from recently dyed yarns in a dyer's booth. Savory scents waft on the breeze as you stride the midway: grilled lamb, chocolate, fried dough. And of course the livestock barns have their own unmistakable aroma of fleece on the hoof, wood shavings, and manure.

To the novice visitor: give yourself the weekend to take in everything. But if something catches your eye, whether it's a rare-breed fleece, an exquisitely dyed skein, a felted rug, or a hand-turned drop spindle, grab it on your first trip through the barn. Chances are very good it won't be there later.

Hand spinners test-drive spindles or look for the perfect festival fleece. Sheep farmers—or prospective sheep farmers—scope out the wares of livestock equipment dealers, the sheep breed pens, or the ewe auction on Saturday morning. Many take in the sheep herding demos or the canine Frisbee competition.

Inside the bazaar, every imaginable sheep product, sheep accessory, or sheep activity competes for your attention—sheep-milk cheese, sheep-themed pottery, sheep objets d'art, sheep jewelry, vintage spinning and textile equipment.

The biggest drawback to attending a booth at a festival is that you don't get to see much beyond your own building. You try to scope things out on trips to the restroom, but there's no time for shopping. The best part of being in a booth is the people who come to find you. I call them our "peeps"; they are loyal, and although the festival is bustling, it is always gratifying to say hello to a customer from Canada or Brazil with whom I've exchanged e-mails for years. Or to see a first skein of hand-spun or a completed pair of socks or a stunning felted piece made from my fiber that's won a ribbon.

In the last hour of the festival on Sunday afternoon, Mike takes over so I can explore. For me, the gem is the one item I know I won't find in a yarn store or online. It may be a one-of-a-kind spindle from Tom Golding. Or a vintage Orenburg lace stole from Galina Khmeleva. Or one of Nan Kennedy's hand-knit sweaters, dyed using seawater, from her Maine seaside flock. There are scads of commercial, mass-produced yarns here, so if I am tempted by yarn, it will be something out of the ordinary. The beauty of "fibering" at a sheep festival is connecting with a shepherd and tapping into the essence of a flock. If that matters to you, don't be afraid to ask vendors if they themselves raised the animals who have produced the fiber in their wares. Ask them, "How was this fiber grown?"

IN A WILD PLACE

Autumn is New England's most visually striking season. At this time of year Patten Hill is a local favorite for autumn outings. Going back years before I ever imagined of owning my own farm here, I visited the Patten from time to time. High Ledges sanctuary was a destination for spring or fall hikes. This six-hundred-plus-acre wildlife preserve is where Mount Massaemet rubs shoulders with its little sister, Patten Hill. Much of its land was once part of the Barnard farm. For naturalists Ellsworth "Dutch" Barnard and his wife, Mary, High Ledges was both their home and their passion. To preserve this pristine place, they donated the land on which the sanctuary sits to the Audubon Society in 1970.

The sanctuary is both remote and accessible, making it a popular day hike for visitors to west Franklin County. Loops of trails named for the sanctuary's native flora and fauna thread their way through a diversity of habitats: swamp, meadow, and woodland. The sanctuary takes its name from the high rocky precipice on the western face, and most hikers come for this view. The trailhead starts at a parking area at the edge of what was once pastureland. Beyond the parking area,

a gravel road well worn by foot traffic wends beneath a mixed-deciduous canopy. The path spans rivulets that feed Spring Brook, part of the Deerfield River watershed. In autumn, sunlight flashes through breaks in the canopy created by falling leaves. The road is carpeted in ocher and yellow. Fallen maple leaves swirl in eddies. The wooded path is hushed.

The trail traverses a long-ago pastured meadow dotted with possum haw and sweet fern. Rusted barbed wire sags on fence posts in repose. Stone walls delineate the edges of pastures gone to brush. Foot-worn paths, trailhead signs, and an occasional bluebird box are the only indications of humans.

My favorite moment to visit this spot is near day's end in autumn. The path stumbles out of the forest, and the Deerfield River valley unfolds before you. From my perch on a granite boulder, the rock face drops away abruptly. I am looking at the tops of trees. On the left, the village of Shelburne Falls appears toy-like. Although the natural features remain the same, the view is altered every year. The advance of civil engineering further underscores the sense of sanctuary that is High Ledges. Resources in the surrounding hills and valley are appropriated for human needs: the water treatment plant, the river belted by hydroelectric dams, a forest incised by a power-line cut, a ridgeline disrupted by a wind turbine. The river winds through the cradle of the Hoosac Range,

home to the Hoosac Rail Tunnel. Mount Greylock, Massachusetts's highest elevation, looms farther west above the Hoosac ridge. The slanted autumn light intensifies the reds and golds of treetops. Shadows delineate every contour. I trace each hill's profile, getting lost in the view and in my own thoughts.

From this spot, the Dutch and Mary Barnard Trail leads me north into the woodland understory. It crosses Spring Brook and then is intersected by the Wolves' Den Trail. This trail leads to a jumbled rock formation known as the Wolves' Den, in what was once part of the "mountain pasture" on the Barnard farm. Local lore says that in the mid-1800s the last known wolf pack in the state of Massachusetts was trapped in the cave and shot by farmers who were irate about losing livestock. As the crow flies, this cave is less than a mile from Springdelle Farm.

Walking the hushed trails of High Ledges is a great reminder of the larger nature just beyond our farm, but I don't have to go into the hills for daily reminders that we live on the edge of a wild place. It's what makes life here interesting and unpredictable.

Back at the farm, a family of porcupines occupies a rocky outcrop near the birch tree lot close to where the boys graze. A few close encounters of the prickly kind have made them wary. Long before our arrival to the neighborhood, a beaver family plugged Spring Brook at the edge of our neighbor Norm's sugar bush. Behind their massive bulwark of sticks and mud, the little pond has become a big pond. When the water level threatens to overtake Patten Road, the highway department fastens a device to the culvert so the water can run beneath the road and into the brook on the other side.

One summer a young bull moose blundered through our fences, freaking out the llamas. In all seasons, the resident gobblers conduct turkey business just outside our door, with little regard for our comings and goings. In late spring, just-roused hungry bears hop the sheep fence, attracted by the anise and molasses in the mineral feed. Once a year, I rescue a giant snapping turtle (perhaps the same one each year) from the net fence near the brook. Occasional reported sightings of catamounts, the local name for the Eastern mountain lion (officially "extinct"), keep us alert. Nature reminds us who was here first.

Coyotes are ubiquitous on the Patten. One family migrates from our dell to the Wheelers' farm, and another one lives in the wooded ledges northwest of our home. I suspect they are close kin. When I'm on lamb watch in spring, the dell coyotes serenade in stereo; the yowls come through both the bedroom window and the barn baby monitor on the nightstand. Yips of juvenile pups join the chorus of adults. I'm grateful that this coyote family has never shown an appetite for lamb.

In fact, despite their reputation for being a nuisance to livestock, we've had no losses from coyotes. Local hunters affably offer to "take care" of our coyotes for us. We politely decline. Our coyotes mind their manners, keeping the rabbit and woodchuck populations under control, we say. We're watchful but feel secure, with our llama bodyguards and fancy woven-wire perimeter fencing to protect our young stock.

Predation is a concern of any keeper of fowl or livestock. Small ruminants have few methods of self-defense: only wariness and flight. It's our responsibility to assess the level of risk and keep them safe.

Our sense of security was shaken to its core one September morning. After the grass growing has slowed down, the flock gets moved more often. The boys were on their last grazing rotation on the sloping pasture behind our house. A

few days of heavy, wet weather stalled hay making. One dripping morning, the clouds sat at ground level, obscuring the view of the pasture.

Mike noticed something odd—a handful of the boys in an agitated huddle at the fence closest to the house. The vibe was strange. Together we drove down to the pasture in the Mule to investigate.

Heavy droplets of rain sagged the net fence. We took it slowly on the slick grass, steering around ledge outcroppings and junipers. Coming over a rise, we saw a snow-like scattering all over the grass. Except it was wool, not snow, strewn about the field. Before we could fully process what we were seeing, we found one of our large Cormo wethers, Phaeton, lifeless, inverted, with his legs splayed in the air. His throat and a portion of his neck had been gnawed to the bone.

Near the base of a tall white pine not fifty feet away, the fence was on the ground. Just outside the fence, we found that our ram Trumpet had also been taken down by the throat. His sheep coat and skin stripped back like a banana peel. His left foreleg almost totally detached. At the pasture's edge, Jack, Trum-

pet's companion, lay still and lifeless. His eyes were open. Unlike the others, he was intact, save for a fatal throat wound.

In stunned silence we walked the field's perimeter while trying to wrap our minds around the situation. Something had run into the fence, and it was down in several places. We were heartsick and shocked by the loss but had to kick into crisis-management mode. Rounding up the traumatized survivors, we found our head count one short. Cosmos, a two-year-old wether, was missing. Mike led the remaining sheep to the pine shelter by the driveway. Armed with a baseball bat and a fire extinguisher, we then set off into the surrounding woods to find our missing sheep. Just below the pasture is a thicket of dense saplings, multiflora rose, and berry brambles. After summer's growth, it's nearly impossible to walk it without a machete, so we didn't get very far. We tried exploring the woods to the south, a mess of deadfall and blowdowns. While Mike returned to the house to alert the neighbors and the authorities, I hopped on the Mule to broaden the search. I didn't know quite what to bring. I kept the fire extinguisher but also grabbed a grain pan and a roll of net fence. Logging roads thread their way

through adjoining woodlots. I covered ground quickly, but not too quickly, as I was also trying to scan the trail for hoofprints, droppings, bits of white.

I eventually ended up in less-familiar territory in our neighbor's pinewood, not far from the beaver pond. The dirt logging trail forked and split several times, and so I veered in different directions. The sunless understory was relatively open, but the trail narrowed and became less obvious. When I reached a dead end, I backtracked, following my own tire tracks. The woods were dripping and dim. One of my sheep was still missing. Three large sheep had been mauled by something that was somewhere here, in these woods. I got a sick, prickly feeling that I was not going to find my ram. I headed back to the house.

Our missing ram was found a day later. The Massachusetts Division of Fisheries and Wildlife sent a biologist to investigate our predator strike. Toting a rifle, he searched the area of the kill site, finding both recent and slightly less recent tracks of a single large canid. We examined the tracks in the mud. The older impressions were compromised by rain, and I would never have spotted them. But the more-recent imprints, less than twenty-four hours old the biologist said, were crisp. They crisscrossed the patch of recently shoveled earth where Norm had laid the boys to rest with his backhoe the day before. The biologist picked up tracks of the pursuit. Our ram had fled downhill, through the woods. The chase ended when the ram became hemmed in by a fence line.

The biologist described the attack as "recreational," since so little had been consumed, hypothesizing that the culprit was a marauding domestic dog. Or possibly a wolf hybrid; there were reports of some released in the Quabbin Reservoir area twenty years ago. Still, so many details did not make sense. What kind of canid would single-handedly take down several large animals in one night? Why was so little consumed? Why attack sheep when smaller game was plentiful at this time of year? How did we not know what was happening right outside our door?

We tightened up security for the rest of the flock. No more rotational grazing on outlying fields this year. At night we brought the lambs into the barn and left floodlights on. The pasture behind the house remained fallow for the rest of the year. We quickly sought an estimate to enclose it with a woven-wire fence. Three weeks later, while in attendance at New York's Sheep and Wool

Festival, we received startling news from home. The canid predator had struck again. This time a neighbor's flock, less than one mile from our farm; a dozen lambs were slain in one night. Many of the details mirrored our attack: all throat wounds and very little consumed, and as at our farm, the predator returned the next night for seconds. This time, however, it was spotted and shot by a neighbor—who at the time thought he was pulling the trigger on a large coyote.

Again the authorities investigated. A visit from another Fisheries and Wildlife biologist revealed this canid was neither dog nor coyote but a gray wolf—a species that had supposedly been eradicated in Massachusetts in the 1840s. The news created quite a stir. The carcass was taken from our neighbor's farm to the University of Massachusetts.

Before the wolf was taken for further study, our neighbor asked if we wanted

to take a look. When we peered into the bed of his truck, we instantly understood why even our large rams had not stood a chance. The wolf's long and powerful legs were made for covering distance. Its broad head and upright ears were densely lined with fur. Bristly gray guard hairs projected over a brown-beige undercoat. Its teeth were clean and healthy. I cupped its front paw in my hand, measuring it against my palm. It was larger. The pads were raspy and the black nails well worn. The wolf was magnificent. Mike and I were struck with both awe and sadness. On the one hand, it was a relief to know that we had likely found the sheep killer. But on the other, I was truly sorry to be making his acquaintance under these circumstances.

Eventually DNA samples were taken at the university and sent to a wildlife forensic lab in Ashland, Oregon. The genetic evidence confirmed that our canid was indeed an Eastern gray wolf, not a wolf hybrid—with no signs of having been in captivity. A male in his prime, he weighed eighty-five pounds and measured more than sixty inches from nose to tail.

The story about the "Shelburne wolf" was picked up by the Associated Press and eventually went national. To this day its appearance here remains an enigma. Speculation was that this male had migrated from Canada, perhaps driven out of its original home by other males. But how did it find its way to the hills of western Massachusetts? There would have been ample food along the way. Why did this healthy animal have a taste for both mutton and lamb when easier prey was abundant? Why did it kill so many at once, consuming so little of its prey? And the biggest nagging question for those of us raising livestock: are there other wolves in Massachusetts?

As with the Eastern mountain lion, the authorities still deny the existence of a wolf population here, calling our experience an anomaly. But reports of wolf sightings continue. Perhaps the wolves are being more furtive. Along with our neighbors, we occasionally hear a distinct howl of a single animal at night. It comes from the wooded slope of Mount Massaemet. The eerie howl stops us in our tracks. I heard it at the barn one evening and recorded it on my iPhone. We and our neighbors, who have lived here much longer than we, agree that it sounds nothing like the yowls and yips of our coyotes. It's a chilling reminder of how much we don't know and how much lies beyond our control.

WILDWOOD CABLED PULLOVER

Designed by Melissa Morgan-Oakes

A mesmerizing Celtic cable is the focal point for Melissa's classic-style pullover. The complex twisting cables are mysterious and alluring, much like a woodland path canopied by writhing oak limbs or a an apple orchard shrouded in wild grapevine. Densely knit wool sweaters are a New England wardrobe staple, and this is a great sweater to wear for a walk in the woods on a brisk fall day.

This sweater is knit with Upland Wool and Alpaca—the softer sister yarn to my Upland Wool and Mohair featured in the Springdelle Jacket. The yarn has a rustic pedigree: the wool comes from our fine-wool crossbred sheep. Many Cocoa daughters, granddaughters, and grandsons contribute to this yarn, and so the yarn's natural colors vary from charcoal to silver to oatmeal, depending on the composition of the flock in any given year. Hand-selected colored alpaca fleece adds warmth to this single-ply yarn, spun at a local woolen mill.

Finished Chest Measurements

34 (38, 42, 46)" (86 [96, 106.5, 117] cm), with an intended 2" (5 cm) of negative ease.

Yarn

- Foxfire Fiber & Designs Upland Wool and Alpaca (80% wool, 20% prime alpaca; 140 yd [128 m]): 11 (12, 13, 14) skeins. Shown in Sumac.

Needles

- One set size US 6 (4 mm) straight needles
- One 16" (40 cm) long circular needle size US 6 (4 mm)

Notions

- Cable needle
- Tapestry needle
- Stitch markers

Gauge

18 sts and 28 rows = 4" (10 cm) in St st. Adjust needle size as necessary to obtain gauge.

FRONT

With straight needles, CO 108 (116, 124, 132) sts. Work p2, k2 rib across first 24 (28, 32, 36) sts, pm, work first row of cable chart across center 60 sts, pm, then work last 24 (28, 32, 36) sts in k2, p2 rib. Keeping in pattern as established, work until piece measures 4" (10 cm) from cast-on edge; end having just worked WS row.

On next row, knit all sts and decrease 7 sts evenly before m, sl m, work the center 60 sts in charted patt, sl m, then knit all sts and decrease another 7 sts evenly spaced to end across the final 24 (28, 32, 36) stitches—94 (102, 110, 118) sts.

Now working first and last 17 (21, 25, 29) sts in St st and keeping center 60 sts in charted patt, work even until piece measures 4¾ (5, 5½, 6)" (11.5 [13, 14, 15] cm) from cast-on edge. End with WS row.

WAIST SHAPING

Row 1 (RS): K1, k2tog, work in established patt to last 3 stitches, ssk, k1.
Rows 2–6: Work 5 rows even in established patt.
Row 7: Repeat Row 1–90 (98, 106, 114) sts.

Work even in established patt until piece measures 9½ (10, 10½ , 11½)" (24 [25.5, 27, 29] cm) from cast-on edge. End with WS row.

BUST SHAPING

Row 1 (RS): K1, M1, work in established patt until 1 st rem, M1, k1.
Rows 2–6: Work 5 rows in patt as established.
Row 7: Repeat Row 1–94 (102, 110, 118) sts.

Work even in patt until piece measures 15½ (16, 16½ , 17½)" (38 [40.5, 42, 44.5] cm) from cast-on edge. End with WS row.

ARMHOLE SHAPING

Bind off 4 sts at beginning of next 2 rows, work rem sts in patt—86 (94, 102, 110) sts.
Row 1: K1, k2tog, work in patt to last 3 sts, ssk, k1.
Row 2: Work even in patt.
Repeat rows 1–2 a total of 5 (7, 9, 10) more times—74 (78, 82, 88) sts rem.

Work in patt until armhole measures 7½ (8, 8½, 9)" (19 [20, 21.5, 23] cm) from bound-off underarm stitches. End with WS row.

NECK SHAPING

Row 1 (RS): In patt, work 27 (29, 31, 34) sts (left shoulder), BO 20 sts, work to end (right shoulder).
Row 2: Work across right shoulder, join second ball of yarn and work across left shoulder. You will now be working each shoulder with separate balls of yarn.
Row 3: Left shoulder: work until 3 sts rem, k2tog, k1; right shoulder: k1, ssk, work to end.
Row 4: Right shoulder: work until 3 sts rem, p2tog through the back loops, p1; right shoulder: p1, p2tog, work to end.

Repeat last 2 rows, decreasing 1 st at neck edge every row until 20 (21, 23, 24) sts remain for each shoulder.

Work as established, without decreasing, until armhole measures 9½ (10, 10½, 11)" (24 [25.5, 27, 29] cm) from armhole BO. End with WS row.

SHOULDER SHAPING

Note: Each shoulder is shaped separately beginning with the left shoulder.

Left Front Shoulder

Row 1 (RS): BO 6 (7, 7, 8) sts, work to end in patt.
Row 2: Work to end.
Row 3: BO 7 (7, 8, 8) sts, work to end.
Row 4: Work to end.
Rows 5: BO rem sts. Cut yarn and pull end through last loop.

Right Front Shoulder

Row 1 (RS): Work in patt.
Row 2: BO 6 (7, 7, 8) sts, work to end.
Row 3: Work to end.
Row 4: BO 7 (7, 8, 8) sts, work to end.
Row 5: Work to end.
Row 6: BO rem sts. Cut yarn and pull end through last loop.

BACK

Work as for Front until armhole shaping is complete. Then continue to work even in established patt until armholes measure 9¼ (9¾ , 10¼, 10¾)" (23.5 [25, 26, 27.5] cm). End with WS row.

NECK SHAPING

Row 1 (RS): Work in established patt across first 20 (21, 23, 24) sts (right shoulder); BO center 34 (36, 36, 40) sts; work across rem 20 (21, 23, 24) sts (left shoulder).
Row 2: Work sts for right shoulder; join a second ball of yarn and work back across left shoulder. Each shoulder is now worked with a separate ball of yarn.

SHOULDER SHAPING

Note: Each shoulder is shaped separately beginning with the right shoulder.

Right Back Shoulder

Row 1 (RS): BO the first 6 (7, 7, 8) sts; work to end in patt as established.
Row 2: Work to end.
Row 3: BO 7 (7, 8, 8) sts, work to end.
Row 4: Work to end.
Row 5: BO remaining sts. Cut yarn and pull end through last loop.

Left Back Shoulder

Row 1 (RS): Work in patt as established.
Row 2: BO 6 (7, 7, 8) sts, work to end.
Row 3: Work to end.
Row 4: BO 7 (7, 8, 8) sts, work to end.
Row 5: Work to end.
Row 6: BO remaining sts. Cut yarn and pull end through last loop.

SLEEVES (MAKES 2)

CO 38 (42, 42, 46) sts. Work in k2, p2 rib for 4" (10 cm). End with WS row. Work 4 rows in St st.

SLEEVE SHAPING

Row 1 (RS): K1, M1, knit to last stitch, M1, k1.
Rows 2–6: Work even in St st.

Work rows 2–6 a total of 13 (12, 14, 13) more times—66 (68, 72, 74) sts.

Work even in established patt until Sleeve measures 18 (18½, 19, 19½)" (45.5 [47, 48, 49.5] cm) from cast-on edge.

BO 4 sts at beginning of next 2 rows—58 (60, 64, 66) sts.

SLEEVE CAP SHAPING

Row 1 (RS): K1, k2tog, knit to last 3 sts, ssk, k1.
Row 2: Purl.
Rows 3–40: Repeat rows 1–2 a total of 19 more times—18 (20, 24, 26) sts.
Row 41: Knit.
Row 42: P1, p2tog tbl, p to last 3 sts, p2tog, p1.
Row 43: Knit.
Row 44: Purl.
Row 45: K1, k2tog, knit to last 3 sts, ssk, k1.
Row 46: Purl.

Work rows 41–46 until 14 sts remain. Bind off remaining sts. Cut yarn and pull end through last loop.

FINISHING

Sew shoulder seams.

With RS facing and using circular needle, beginning at back right neck edge, pick up and knit 34 (36, 36, 40) sts across back neck; pick up and knit 7 (8,

8, 10) sts along left shoulder and sloped neck edge; pick up and knit 20 sts across front neck edge; and pick up and knit 7 (8, 8, 10) sts along right sloped neck edge and shoulder—68 (72, 72, 80) sts.

Work in the round in k2, p2 rib for 2½" (6 cm). BO loosely. Cut yarn and pull end through last loop.

With tapestry needle, sew Sleeves into armholes, easing to fit; sew sides from cast-on edges to underarm. Block prior to wearing. Enjoy!

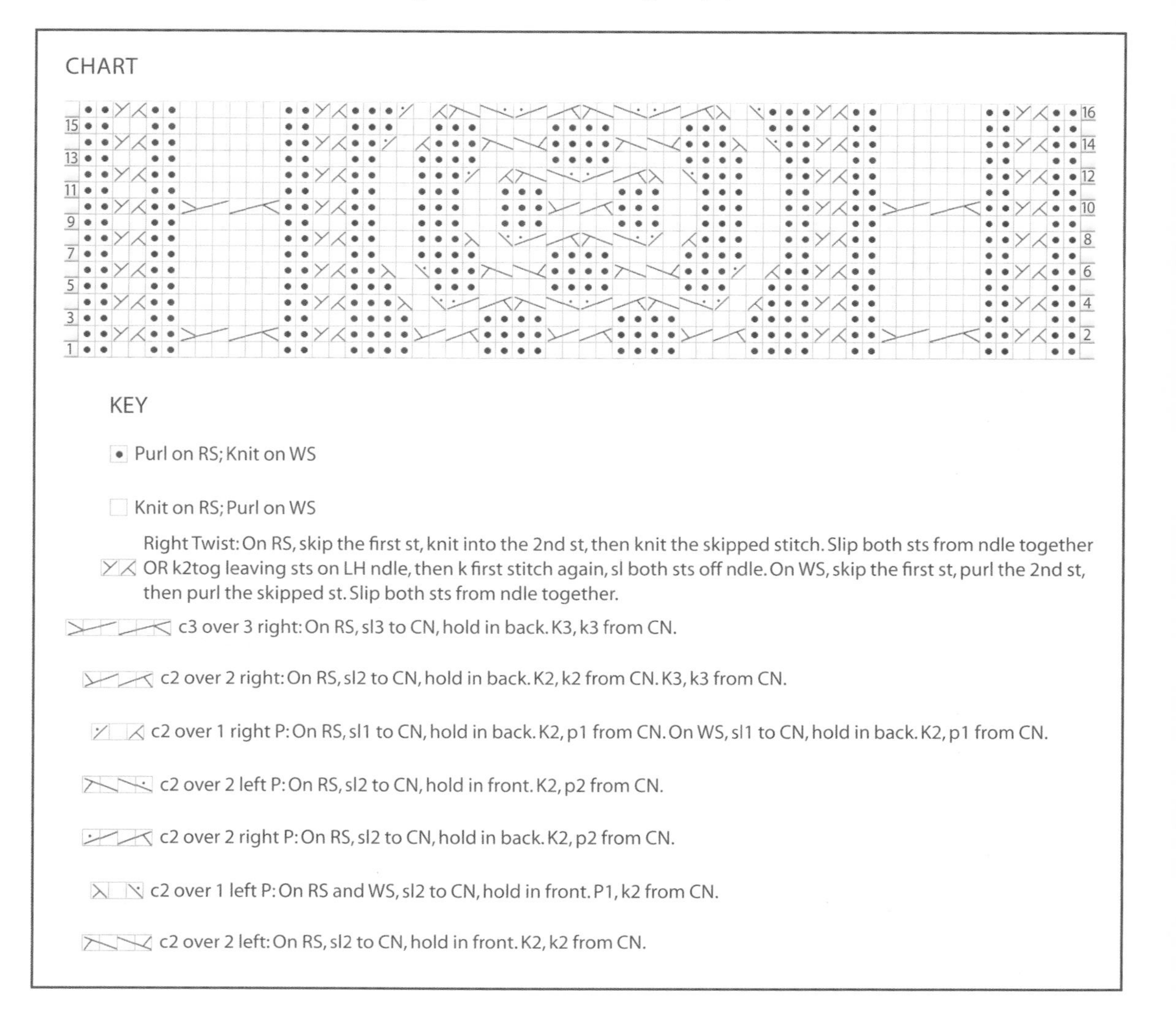

US

WINTER

Winter often arrives before fall has officially departed. It's not unusual for the sheep to scrape through a crystal rind to get to the grass on their last graze of the season. There is a feeling of contraction as winter knits the farm into a tight enclave. We gather up the flock to the barn paddocks, and the barns are once again the hub of activity. We collect fencing and field equipment. The farm gets tucked in. Winter isn't exactly a time of hibernation but, rather, a time of suspended activity. As the roads glaze and our fields eventually fill with snow, focus shifts inward. After evening chores and dinner, I settle into a comfy chair with a lapful of yarn and a pair of needles; the flock hunkers down within the fold.

GATHERING THE FLOCK

WINTER'S ARRIVAL is unpredictable. In most years, winter knocks at our door just before Thanksgiving. But it has also ambushed us with a foot of snow as early as Halloween. When I find an inch of ice capping the water buckets and have to fish it out with my bare hand, I realize it's time to put the farm to bed. I start a mental list: set up the electric heated buckets; keep a pair of insulated work gloves in the car; collect and drain the hoses running to the paddock; round up buckets, stock tanks, and troughs in the field. A series of frosts hardens the ground, creating minor hazards. Divots from hooves, hummocks of frozen grass, even llama dung piles are potential ankle-wrenching moguls. We eke out grazing as long as possible, as each day dining in the fields conserves approximately 350 pounds of hay in the barn. We ask the sheep to march a bit farther afield to clean up what remains in the mowings and in the places that were too wet to mow in the summer. But the grass dwindles and, after several hard frosts, goes dormant.

In what felt like a lucky mild December one year, we managed to extend grazing right up to the solstice. But a sudden ice storm caught us off guard. We still had sheep in the field. Six rolls of net fence, one thousand feet, had yet to come down. Rain turned to freezing rain, followed by a cold blast. The spikes at the bottom of each post (one every twelve feet) froze fast in the ground. Using a screwdriver as a chisel with a hammer, we chipped every stake out of the frozen turf before we could roll that fence up and put it away. That year taught us a lesson. If a December forecast predicts rain in the valley, it means freezing rain, sleet, or snow in the hill towns. We never again waited till the last minute to take down fencing.

As courtship season ends, we rearrange the flock, settling them into their winter accommodations. With reluctance, Parsley, Teaberry, and Chai take leave of their ladies. They join the rest of the boys, whom we've trailered down to the lower farm from the hilltop. The carriage barn, stacked to the rafters with square bales, is their winter home. The bred ewes' winter enclave is the south end of the dairy barn. Crackerjack joins them there, where they will birth their lambs in April. The open barn across the street houses the goats, llamas, Caitlyn and Sol, and the rest of the flock: There's a group of unbred ewes—the ones who won't be lambing this year—at the west end with a pair of llamas for housemates. A separate pen houses the grannies—Cocoa, Charlotte, Buttercup, and Pansy—who, as our most senior flock members, receive special rations. Each pen requires water, hay, and minerals. Smaller groups let us attend to the individual needs of different animals. But the more groups, the more work there is for us.

The young replacement ewe lambs we are keeping from spring reside in the pen closest to the road at the east end of the open barn. Because they're smaller and still growing, we don't force them to compete for feeder space with the pushy, larger girls.

The number of ewe lambs we keep in the flock varies, depending on the year. We try to keep a daughter (or granddaughter) to continue the lineage of each original herd matriarch. But the summer's hay harvest determines how many sheep our farm can support through the winter months. To reduce our flock, several ewe lambs have been sold as breeding stock to other farms, and sometimes we get calls from farmers looking for a mature ram lamb to run with their ewes. Then, chances are, they'll put him in the freezer eventually. Our other ram lambs, and occasionally some ewes as well, are also spoken for. We have a following of customers, some local, many from urban areas, who call me every year looking for freezer lamb. We reserve several lambs for our local food co-op and two for a local restaurant that goes out of its way to support local farms.

People are sometimes surprised when I tell them that each year we raise some lambs for meat. There seems to be an assumption that if you love animals and treat them well, you cannot bear to send them to slaughter. Mike and I love all of our sheep and take the responsibility of caring for them seriously. We would do the flock a disservice if we held on to all of them. We are stewards of our flock and land, and our farm is a finite resource. As with all farms, its carrying capacity varies from year to year. Well-managed land provides enough food for our livestock. Our livestock provide us with warm fiber and healthy food. We see them into the world and respect their presence by giving them names. We treat all of them compassionately, as sentient beings, regardless of gender or how long they may stay.

But each fall some lambs take their leave. Either Holly or I, or sometimes Fred, will take them to the slaughterhouse an hour away. We load them calmly into the truck, or into a rented van if there's a large group going, or into the back of my Highlander if it's just two or three. I take them the day of slaughter if possible, to spare them a night in a strange place. There is always a pang of sadness and regret at seeing them go. That's appropriate, and I don't think it will ever change. If it did, it would be time for me to stop what I do.

A week after the lambs are dropped off, my customers come to the farm with large coolers to pick up their meat. Some of them are old friends. Some I see only once a year, at this time: a retired professor and his wife from Boston; a minister and his wife come with their young daughter. Coming to the farm is important to them. They want to see how the sheep have been raised and to know who has raised their food. It's an honor to do this work for them. Several will also make a stop at the Wheelers' farm to pick up beef.

As pasture forage dwindles, the flock becomes keen for our morning arrival. Acute declines in temperature sharpen appetites. The appearance of my car at the gate causes ripples of excitement. Before I can unfasten the latch, the girls vie for position at the feeder. As the seasons shift, dining migrates from field to feeder. Until now, they've been able to march out to pasture at daybreak or whenever they're hungry. As we transition from pasture to hay, they rely heavily on our punctuality. If we're running behind schedule for feeding time, they let us know.

Feeding indoors presents a different set of challenges in sheep and wool management than raising sheep on pasture. Fresh flakes of second cutting—timothy and clover—smell like summer. Sixty-four feet of feeders run down the center of the open barn. A plywood partition separates the sheep side of the barn, where they hang out, from the people side, where we store grain and hay. I start at the barn's west end, tossing hay flakes into the bunks from behind the plywood wall in back of the feeders. It's the only way to feed them without getting trampled by hooves and without dropping hay over their backs. The ewes use their shoulders to carve out space at the feeders. They bite off more than they can chew, nosing greedily into the slots of the hayrack, tearing out lusty mouthfuls with their front teeth. Because they don't want to miss a thing, they look to the right and to the left as they work a mouthful of hay to their grinding teeth at the back of their mouth. A shower of chaff and debris litters the necks and shoulders of their barn mates.

Watering the sheep in winter presents its own set of challenges. The frost-free spigot is cranky and, on very cold mornings, despite its name, not fully free of frost. Its pump handle takes persuasion before delivering a slow but steady stream to fill buckets. At the north end of the barn, I check the valve on the

Nelson waterer, a round device about fifteen inches in diameter set into the concrete floor. Inside its metal housing, a small stainless bowl sits on a balance arm. When the balance arm detects a drop in the bowl's water level, it refills itself automatically via a tiny spigot. A small electric heating element keeps the spigot and bowl ice-free. When we installed the Nelson, we placed it strategically where it could be shared by sheep in two adjacent pens. The grannies have access from one side. The young ewes and llamas can drink from the other. Twice a day it takes many jugs filled in the milk-room sink to replenish the boys' buckets in the carriage barn, a hundred feet away. A four-gallon jug weighs more than thirty-three pounds. We lug them in all kinds of weather up the incline to the carriage barn. Another round of jugs gets carted down the aisle of the dairy to the mothers. Mike is a good sport about lugging water while I dole out hay. In winter, we divide tasks and waste no time.

We develop a new routine and fall into a rhythm as daylight minutes retreat earlier each afternoon, counting down to the solstice. It works fine, until nature throws a curveball.

ICE STORM

IN READYING FOR WINTER, we gear up for snow and wind. We fit our three-quarter-ton truck with an eight-foot plow and studded tires. The last person to drive it must top off the gas. We park it facing outward in the high drive of the barn. Mike's John Deere tractor swaps its front-end loader for a six-foot snow thrower. Its steel auger can propel the snow twelve feet off to the side—keeping our half-mile driveway from getting socked in. In past winters, the wind has been strong enough to wiggle fifteen-feet-tall barn doors from their latches. Attached to the buildings only by the track at the top, the wooden doors swing like giant sails outward into the paddock. We batten them down with heavy storm keepers—two-by-six beams that slide into iron brackets on the inside of the pair of doors on the windward face of the carriage barn. Safety chains secure the westward-facing gates.

And yet we were caught flat-footed when two weeks before Christmas 2008 one of the worst ice storms in decades assaulted the Northeast. Snowfall was light in the weeks before the storm, and a deep frost had yet to settle in. Warnings that we were in for a wintry mix gave us time to prepare the farm, giving extra hay to the sheep, topping off buckets, securing latches on barn doors, and tightening up gates.

We half slept that night to the sound of wind howling in the chimney and icy shards pelting the panes. Just after midnight the generator that powers the house in the event of outages kicked on. The telephone rang: our security company alerting us that the power was also out at the lower farm. I immediately started thinking about the sheep. Without power, there would be no water. The extra buckets put out at dinner would be ice blocks by now. Maybe the large heated buckets would stay unfrozen until the power was restored in the morning.

Sharp explosive sounds like rifle shots woke us before dawn, treetops snapping in the woods. The cracking was followed by the sound of crashing branches and percussive thuds as tree after tree toppled. Waiting for daylight, we sipped coffee, grimly listening to the popping woods and the thrumming generator.

At first light the view from the front window was not encouraging. A large

sugar maple lay shattered in the front garden, blocking the driveway. The cathedral of hemlock and white pine to the west was veiled in ice. Bowing birches formed crystal curtains across the drive. Inch-thick ice sheathed every limb and twig. Grasses, shrubs, stone walls, lampposts were all glistening and encased. The sickening sound of trees toppling continued and made us anxious to get down to the farm. I filled plastic milk jugs with water for the animals and loaded them into Mike's SUV.

Getting to the lower farm was not so simple. Although there was almost no snow and the driveway wasn't terribly slick, we stopped every fifty yards to shake the ice from branches of the bowing birches, to make them spring out of the way. Where the trees were too bowed to spring back, we were forced to detour off the road. Around the bend, the track was littered with broken pine limbs. We dragged whatever we could out of the way, but for heavy limbs we needed a tractor with chains. Again we resorted to off-road maneuvers.

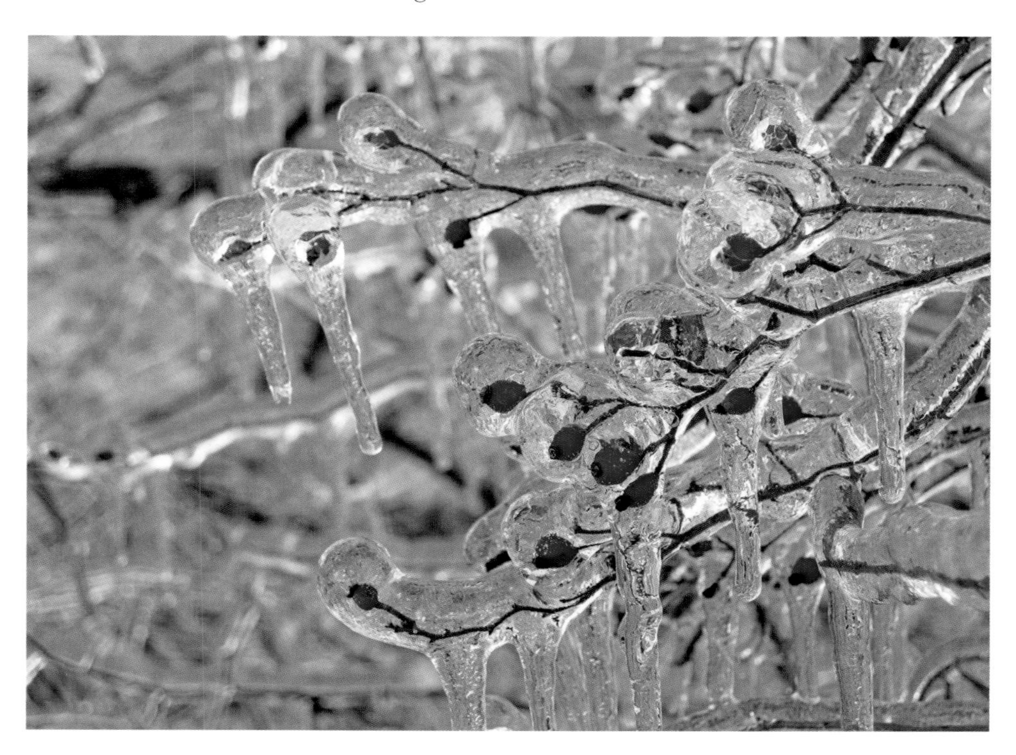

It took nearly an hour to claw our way the half mile down the driveway. At the gate we were barred again by a tree that had fallen, caving in the gatepost lantern on the right. Its branches were hung up on the iron gate. As Mike hopped out to drag away the branches, a heavy limb from another tree came crashing down, taking out the gatepost lantern on the left just a few feet away, sending shards of glass skittering across the icy drive.

At the road, toppled trunks and snapped limbs dangled precariously from telephone wires; power lines lay on the ground. But the neighborhood had mobilized. We heard the sounds of tractors and chain saws.

Our apprehension rose as we crept our way downhill toward the farm. The upper Patten was a mess. Everywhere, the weight of ice had torn large trees right off at the crown. Before we reached our barns, we saw the storm's toll on our pastures. Tall white pines delimbed by ice. The trunks of ancient sugar maples split in two. In one field at least one hundred feet of fence was staved in by tree trunks.

While the woodland was in a state of havoc, the farm complex itself was amazingly unscathed. Trees had not landed on roofs. Gates and fences immediately surrounding buildings were intact. The sheep were alert but unfazed, blinking in the brightness as grayness gave way to sunlight bouncing off the crystalline landscape. As long as we show up twice a day with hay and fresh water, all is well in their world. With the power out at the farm, it took two more trips from the house to water the entire flock that morning.

We banged ice from buckets, inspected the water pipes in the milk room, freed up frozen gate latches and doors. It took days to fully realize the storm's impact on our farm. Elevated areas took the brunt of the damage. The face of the hillside above the barns had been ravaged. Dozens of trees were snapped at the top, leaving exposed heartwood and uneven gashes in the tree line. Our wire fences had been hammered by fallen trees and limbs. Where our upper pasture abutted Norm's woodlot, the fence line was buried in shattered limbs and branches. Behind our home, our newly installed, post-wolf-attack fence was on the ground.

Ironically, the week before the storm I had hired an arborist to remove a weak limb on a magnificent old maple beside the fence at the pasture's edge. Now its massive trunk had split in half, wiping out a good-sized swath of fence.

Although stunned by the overnight devastation of the landscape, we realized we were fortunate. No one was hurt. Norm came with a tractor and heavy chains to make our driveway passable. Neighbors with chain saws attacked the mess on Patten Road. Eventually, the electric company repaired the power line that dangled near the end of our driveway and restored power. Our perimeter fences were a shambles, and it would take weeks to get them cleared.

For weeks convoys of bright-orange Asplundh trucks with cherry pickers and chippers descended on the country to clear trees and power lines from roads. Elsewhere in New England the higher elevations bore the brunt of the storm. Utility companies worked around the clock. To this day we are reminded of this storm. Jagged trunks and piles of limbs along the wood trail are cautionary reminders. Each year we clean up a bit more of the damage to our woodlands.

WINTER SOLSTICE

DESPITE UNPREDICTABLE WEATHER, winter on the farm is beautiful. In the days leading to the solstice, the sheep are doing what sheep do best in winter: eating prodigious amounts of hay, depositing massive poop piles for us to pick out of the straw, and growing wool. Between holiday errands and shortening days, it gets harder to arrive for evening chores before daylight exits. I work the gate by feel and slide the barn door to the right.

On the solstice I arrive before dusk. In the soft transitional light I pop a bale onto the sled, cut the twine, and haul the bale out the back door into the paddock. The sled glides easily through fresh powder. Forty pounds of fragrant dried grass is enough incentive for hungry sheep. Dominant ewes and a goat or two lead the way. Everyone spills out of the barn, tailing along like a comet. I drag the sled in an arc around the field, dropping hay flakes every few steps onto clean snow. The snow is barely up to the llama's knees, but for the hoggets, it's chest deep. Like little snowplows, they barrel through. The flock fans out in groups of three or four to munch on dried grass in the snow-covered field.

When the snow is clean, we can get away with feeding on the ground. Wool growth is heaviest in the weeks leading up to winter. As fleeces burgeon, the ewes widen in girth, making space tight at the barn feeders. In the coming weeks there will be many days when we'll be forced to feed outdoors.

I hide treats in my pockets. A clementine for Cocoa; bread for the goats, who shove their noses right into my pockets and, once they've emptied them, insist on eating hay directly from the sled.

While the flock cleans up every bit of hay, I take care of housekeeping: scooping up mounds of droppings with a shovel and adding fresh bedding straw to the pens—though on a still, clear night like this, the sheep may decide to camp under the stars, making sheep-sized hollows in the snow.

We are generally home for the December holidays and the week immediately before and after, as it's not easy to get farm help on Christmas Day. The week before Christmas in 2010, Mike and I took a little detour on our way to an appointment in Boston. A day earlier I had received an e-mail from the director of the Animal Rescue League of Boston's large-animal facility in Dedham. It was a desperate plea to the fiber community. A seven-year-old llama gelding needed a home.

This llama came with baggage. According to the e-mail, he was brought to the shelter in September after escaping from his previous home somewhere in Massachusetts and running loose for *two months*. It took a team of concerned people to capture him and get him to the rescue barn.

Until the moment of the e-mail's arrival, Mike and I had exactly zero plans to adopt anything; we certainly weren't looking for a llama, with or without baggage. But something about his story haunted us and made us decide to stop at the animal shelter on our way to Boston to meet him.

We learned more about his prefugitive history. According to the shelter's director, his prior owners had acquired him at auction and may not have been equipped to house a llama. He got loose. He stayed loose. Once he was caught, they didn't want him back. At the shelter he was tested for various diseases (all negative), inoculated, castrated, microchipped, and placed in a paddock with three sheep adjacent to a paddock of horses and a pig.

Having got stuck in rush hour traffic near Boston, Mike and I arrived very late in the afternoon. By the time Lisa, the shelter's director, led us to the paddock, dusk had settled. We could barely see the sable-brown llama standing at the other side of the paddock with three sheep circled about his legs. The white dash on his face and neck were his only distinctly visible features. About the only things we could gather were that he was very people shy (they were unable to catch him so we could have a closer look), he was fast, and he didn't seem to mind having sheep clustered around his legs. That was important. We left the shelter without making any promises other than to think about it. But less than thirty minutes from the shelter, while (again) sitting in traffic, I called the shelter. Mike and I would give this llama a home.

The following week, on the winter solstice, Andy and I left the farm at 5:45 A.M. with the horse trailer hitched to the truck. I had been up at three A.M. that morning to view the lunar eclipse. After picking up the llama, we drove the Mass Pike home in a blizzard, with cars sliding in all directions. I was glad to have Andy behind the wheel and for Andy's firm but patient handling of a very nervous llama.

The shelter had warned us that he was a spitter. But he hadn't spit at me or

Andy. Nor did he flinch, spit, or kick at the curious sheep who excitedly crowded at the fence to get a look at the newcomer when we unloaded him at our barn. He didn't spit at Caitlyn, our female llama, who came over to inspect him.

While stressed and bewildered after arriving at yet another new place after a two-hour trailer ride, he was more curious than fearful. We placed him with last year's lambs, who were used to having a llama companion and least likely to give him attitude. This move also freed up Crackerjack to be with our expectant ewes—we had already moved him over to the birthing barn across the street. For the lambs, it was love at first sight. They gathered around the llama's legs with welcoming sniffs, tilting their woolly heads way back for a good look

at the new tall, dark, and handsome guy. Andy and I were in awe of how mellow this llama was, given everything he had been through.

Any adoption is a leap of faith. You commit and hope for the best. We hoped that in time he would learn to trust us. The lambs and Caitlyn modeled good barn behavior. For the first few weeks, we gave him space and let him get used to our voices and routines. And we made doubly sure all doors and gates were secure.

We named him Sol, since he arrived at Springdelle on the winter solstice. To this day, he is still wary and enigmatic. We keep our expectations low for him and are careful to respect his space. He follows Caitlyn and his flock mates, preferring their company to ours.

SOLSTICE HAT

Designed by Christa Giles

A wool cap is a comforting essential, even on mild winter days. Christa's design for a simple winter cap plays with color and texture. The shaping for the cap, which tapers toward the back of the head, makes it wearable beneath a hood or fit comfortably when bundled in a high-collared parka.

Cormo wool makes this hat clingy without being hair flattening. The luster of silk defines the stitches.

Finished Measurements

18¼" (46.5 cm) circumference unstretched, will stretch up to 22" (56 cm)

Yarn

- Foxfire Fiber & Designs Cormo Silk Alpaca (70% Cormo, 20% alpaca, 10% silk); 190 yd [174 m]: 1 skein main color (MC), 1 skein contrast color (CC). Shown in Hyssop (MC) and Katydid (CC). Sample used approx 132 yd (120 m) in MC, 22 yd (20 m) in CC.
- Small amount of waste yarn in same or lighter weight

Needles

- One 16" (40 cm) long circular needle size US 7 (4.5 mm)
- One set of five double-pointed needles size US 7 (4.5 mm)

Notions

- Stitch markers
- Tapestry needle or bodkin

Gauge

21 sts and 24 rows = 4" (10 cm) in St st. Adjust needle size as necessary to obtain gauge.

Note: This hat is worked from the top down, with the piping knitted in as you go.

HAT

With dpns and waste yarn, CO 8 sts, redistribute over 4 dpns and join, taking care not to twist. Knit one rnd.

Change to MC, leaving a 6" (15 cm) tail at the beginning to be used later to cinch the top of the hat.

Rnd 1: (K1, yo, pm) eight times—16 sts.
Rnd 2: (K1, k1 tbl, sl m) eight times.
Rnd 3: (Knit to m, yo, sl m) eight times (8 sts added).
Rnd 4: (Knit to previous round's yo, k1 tbl, sl m) eight times.

Rep rnds 3–4 until there are 11 stitches in each section between markers, switching to circular needle when working the final rep of rnd 4 (or sooner if you prefer)—88 sts.

PIPING

Do not cut MC. Knit next 3 rnds with CC, removing all markers except start-of-round marker. Cut CC, leaving sufficient tail for darning in.

Using MC, * reach behind to purl side and use tip of right-hand needle to pick up loop of next stitch 3 rows below (the bump of CC that interlocks with the MC stitches 4 rows below), place on left-hand needle, and knit together with next stitch; repeat from * to end of rnd.

TEXTURE PATTERN

Rnd 1: *K11, yo; repeat from * to end of rnd—96 sts.
Rnd 2: *K2, p2, k2, p2, k2, p1, p1 tbl; repeat from * to end of rnd.
Rnd 3: *K2, p2; repeat from * to end of rnd.
Rnds 4 and 5: Knit.
Rnds 6 and 7: *K2, p2; repeat from * to end of rnd.
Rnd 8: Knit.
Repeat rnds 4–7 four more times.
Knit one rnd, then repeat piping.

RIBBING

Note: Wraps do not need to be picked up on following rows, as they will blend into the ribbing pattern. The next section is worked with short rows.

Row 1(RS): (P1, k1) 31 times, w&t.
Row 2: (P1, k1) 12 times, p1, w&t.
Row 3: K1, (p1, k1) 16 times, w&t.
Row 4: (P1, k1) 20 times, p1, w&t.
Row 5: K1, (p1, k1) 24 times, w&t.
Row 6: (P1, k1) 28 times, p1, w&t.
Row 7: K1, (p1, k1) 30 times, w&t.
Row 8: (P1, k1) 32 times, p1, w&t.
Row 9: K1, (p1, k1) 34 times, w&t.
Row 10: (P1, k1) 36 times, p1, w&t .

Row 11: K1, (p1, k1) 38 times, w&t.
Row 12: (P1, k1) 40 times, p1, w&t.
Row 13: K1, (p1, k1) 42 times, w&t.
Row 14: (P1, k1) 44 times, p1, w&t.
Row 15: K1, *p1, k1; repeat from * to end.

You are now working in rnds. Work (p1, k1) for four rnds, then bind off loosely in pattern. Carefully remove the waste yarn from the top of the hat and weave the MC tail through the live loops to cinch closed.

Weave in ends, block, and enjoy!

SNOWSHOE SHEPHERD

SHEEP ARE REMARKABLY WELL ADAPTED for snow. Their manure piles tell me where they've spent the night. Sometimes they'll bed outside in light snowfall or on windless nights. Tucking their legs beneath their bodies, they fold themselves into depressions in the snow. Fully insulated by wool, they may accumulate an inch of snow on their backs.

Wind drives them into the barn. The open barn has the most exposure, as the south wall is open to the weather. The sheep pens are flooded with sun on mild days, but if we don't close the doors against the prevailing wind at the western gable ends, the barn becomes a wind tunnel.

A sixteen-foot galvanized-steel-panel gate is like a giant sail. We use this gate every day to feed the flock in the open barn. When the wind barrels over the hilltop in thirty-five-mile-per-hour gusts, we time the opening of the gate between gusts.

I will take a cold winter day over a hot summer day. Cold encourages you to move quickly and be efficient. I dress in layers according to conditions: wool long underwear, sweater, flannel-lined Carhartt bibs, a Gore-Tex shell, a fleece face mask, waterproof work gloves. Someone once told me there's no such thing as bad weather, just inappropriate clothing. Despite the punishing weather I still prefer winter's coldest days to summer's hottest days. On the coldest days, I can always add another layer. On the hottest summer days, you can get away with removing only so much.

As much as we hustle through chores, some tasks always take a certain amount of time. There is no hurrying the rate at which the spigot fills the buckets. I use the pauses in the action to visit and study my flock. While waiting for buckets to fill I look down the line of woolly bodies to see if anyone needs a larger coat. I also look for signs of trouble.

Several years ago Amy, a black Leicester ewe, developed frostbite. I noticed one day when she was flicking her ears repeatedly that the tips were covered in scaly white patches. Holly fitted her with a pair of earmuffs, but Amy refused to wear them. We moved her to the dairy barn to make her more comfortable. One of her ears has a permanent notch from frostbite, and to this day her ears seem especially vulnerable to cold.

The granny ewes struggle with winter. In her last winter, Charlotte developed a chest rattle. We moved her to a pen in the heated milk room and started her on a course of antibiotics. After a week of TLC she was strong enough to rejoin her flock mates in the barn.

Beyond the health concerns of our flock, we also need to watch our food stores. Come late January in a heavy winter we burn through hay. By this point in the season we have gone through the caches of square bales in the carriage barn and the open barn. From the north end of the hayloft in the dairy barn, I toss bales into the bed of the truck and ferry them to the other buildings. The once seemingly vast stockpile in the loft diminishes at an alarming rate. On clear nights when temperatures dip into the single digits, I add an extra half bale to the feeders and toss in a few handfuls of corn for extra calories. The sheep also conserve energy. By day they bask like lizards in the sun, and they hunker down in the straw after dinner each evening.

The key to winter feeding is finding a balance for each group, feeding them just enough to maintain condition but not so much that the sheep get fat or wasteful. For each of the various subflocks, we make notes on a feeding chart that hangs in the milk room. A glance shows the number of sheep in any given group and the amount of hay that group should receive per day. Sticking to the chart is important. I used the same numbers back in the fall to calculate how much hay to put up. Nonetheless, depending on the severity of the winter, I get alarmed. If we get careless with feeding, there may not be enough to get us through the season. So no matter how hard the sheep try to persuade us otherwise, we stick to the rations for each group.

Years ago Norm gave me a helpful tip: if you still have half your hay by Groundhog Day, you should make it until spring. In some years, by the second of February I am in the loft, counting up the remaining bales.

In January, when a sharp, icy rind caps shin-deep snow, the sheep are reluctant to leave the barn. I strap on snowshoes and break a trail through the crust, stomping back and forth to tramp the ice into pieces and compact the snow. Sheep don't mind a crunchy path. With the flock at my heels, I tow the hay sled out into the pasture for breakfast.

The wind sails over the tree line from the west, polishing the fields to a marzipan sheen. Stark and devoid of color, the bones and contours of the land are legible. From the plateau behind the open barn where the sheep are dining, I get a glimpse of the terrain of the laurel swamp. In summer you can't see into the slope, let alone walk it. It's a tangle of saplings, sumac, bittersweet vine, and multiflora rose. Before our arrival, Holsteins pastured this land, but since mountain laurel is toxic to sheep, this area is now off-limits to the flock. Studying it from below, I often think that if we could get the slope back under control, we could eventually form contiguous grazing paddocks between the lower farm and the large pasture behind our house. It would save us the task of trailering the boys back and forth in spring and fall.

A good snowpack on a windless winter morning is the perfect time to investigate. I started in the field behind the house and work my way downhill to the farm. The otherwise impenetrable snarl of roses and wild berry brambles is flattened with snow and ice, forming a springy base on which to snowshoe. The crust is strong enough to support the weight of deer. Following their tracks through the maze of saplings, I find clues about the farm's history. I discover a spring-fed pond in a hollow just below the house. It must have been used to provide water for the cows in this field. Farther downhill I find a stone wall, the upper reach of which I see from the bottom of the hill behind the open barn. A second stone wall runs parallel to it about four hundred feet away. At one time, all of those stones were in the pasture. I wonder who hauled them from the field and laid them there and how long ago. It looks like maybe there was a road that ran between the two stone walls—we've often thought how useful it would be to be able to drive from top to bottom without cutting through the hay field. At the lower end of the slope the ground is lumpy and dotted with

laurel. Remnants of an old fence run through the brush parallel to our sheep fence. Drainage here is poor; I crunch through the ice on standing water. The field peters out where it intersects a brook. The brook begins at an old stone-laid well at the toe of the hill in a stand of maples. This was once the water supply for the farm. A fallen limb has punched a hole in the roof of the small well house. I peer inside. Slightly downhill from the well house, there's a small man-made pond. I mentally map this region of the farm for future reference.

Snowshoe treks reveal more about the farm's woodland residents. The coyotes leave scat in the middle of our driveway where they cross on a regular basis. I pause to examine a fresh deposit. Coyote scat and fox scat are similar; the diameter and large bone fragments in this specimen say "coyote." In winter, the area between the apple trees is their most popular crossing. I want to see where they go and set off into the mixed hardwoods north of the driveway one afternoon.

Snowshoeing here is easier than in the swamp. The woodland canopy has shaded out the rough undergrowth, but still I have to step carefully over fallen trees. Not far into the woods, there are signs of more coyote traffic. The trail I'm following is intersected at various points by other tracks. In places the coyote I'm tracking has stopped to urinate, leaving rusty drops of blood on the snow. The tracks lead into a ravine, and I find evidence of a chase with a rabbit.

As I approach a craggy seam of ledge running north–south, I wonder if the trail will lead me to a den. The prints go around the outcrop. I spy crevices that look like they could house a critter, but my coyote's trail continues up the ravine, over the ledge to the north. Eventually it leads me to where my farm meets the town's watershed property. I can understand why the watershed property would make a good coyote haven. The land is well posted on all perimeters, so there is almost zero human traffic. The parcel abuts the High Ledges sanctuary land, increasing the protected acreage. A stone wall and four strands of barbed wire separate my woodlot from the watershed woods. My coyote has found a break in the stones and slipped beneath the barbed wire. Tufts of wiry gray fur are snagged on the fence. I pull some of them off. As I finger them, my mind automatically contemplates the properties of coyote fiber and what type of yarn it would make. I stuff them into my pocket. That's as far as I'll go. The coyote's final destination remains a mystery. Its tracks continue into the woods beyond.

DIP-DYED YARN

Dip-dyeing is a shortcut for creating a yarn with color segments, without going to the trouble of painting each skein.

Materials

- 2 skeins sport weight yarn (I used 60% wool from Cocoa, my first ewe, and 40% alpaca)
- Cushing's Perfection dyes in 3 colors (Plum, Navy, Mahogany)
- White vinegar

Equipment

- Basin of warm water
- Stainless or enamel stockpot (10 quart size works well)
- Propane burner or cooktop
- Slotted spoon
- Dowel

Before getting started, please read and follow the Guidelines for Safely Dyeing Yarn and Fiber on page 305.

Presoak the skeins in a basin of warm water for 30 minutes.

Mix the dye powder into solution. Cushing's dyes come in little packets. Mix each packet with 4 cups of boiling water. (You won't use all of the dye for this project, but you can store unused dye stock in plastic milk jugs for up to six months.)

Remove the skeins from the presoak and wring out excess water.

Fill a stockpot one-third full with water and place on burner set on high. Add ½ cup vinegar and ½ cup of your first dye solution to pot. Stir. Just before the water comes to a boil, dip one-third of each yarn skein into the dye bath. Wrap the remainder of the skeins around a metal spoon or dowel to keep it out of the water. Let the yarn simmer in the first color for 30 minutes or until all the color has been absorbed. Adjust the temperature to keep the pot just under a boil. Carefully lift skeins from the pot.

Add ½ cup of next color to the pot and stir. Reposition the skeins over the pot. Submerge the next third of the skeins into the second color. Be sure to not submerge the section dyed in first batch of dye. Suspend the skeins wrapped around the spoon over the pot. Simmer for 30 minutes or until the yarn has absorbed all color from water. Lift the skeins from the pot.

Add ½ cup of the third color and stir. Position the skeins so the remaining third dangles from the spoon into the pot. Simmer for 30 minutes until the yarn takes the color. When the water is clear or nearly clear, submerge the entire skeins in the pot. Let simmer for another 30 minutes. Allow the skeins to cool in the pot.

Rinse skeins in basin and hang to dry.

MAX0104
0269

FULL SNOW MOON COWL AND MITTS

Designed by Barbara Giguere

The full snow moon comes in February, as winter turns a corner. This cozy cowl and the matching fingerless mitts work better than a scarf and mittens for winter-morning chores inside the barn. There are no scarf tails to get caught on feeders; fingers are free to work the latches on pens. Barbara's cowl and mitts translate equally well for running errands or a walk to the mailbox.

This project is knit from the wool of Cocoa, my very first ewe, who now enters her sixteenth winter.

Finished Measurements

Cowl: 20½" (52 cm) around at the base and 7" (17.5 cm) high

Mitts: 6½" (16.5 cm) around at the cuff

Yarn

Cowl: Hand-spun sport weight yarn (250 yd [228.5 m]): 1 skein. Shown in yarn that was hand-spun using 60% wool, 40% alpaca then dyed using the previously described dip-dye method.

Mitts: Hand-spun sport weight yarn (250 yd [228.5 m]): 1 skein. Shown in yarn that was hand-spun using 60% wool, 40% alpaca then dyed using the previously described dip-dye method.

Needles

Cowl: One set 16" (40 cm) long circular needles size US 9 (5.5 mm)

Mitts: One set size US 5 (3.74 mm) straight needles

One set double-pointed needles size US 5 (3.75 mm)

Notions

- Cable needle
- Stitch markers

Gauge

Cowl: About 12 sts = 4" (10 cm). Adjust needle size as necessary to obtain gauge.
Mitts: 20 sts = 4" (10 cm). Adjust needle size as necessary to obtain gauge.
Note: You can adjust size of the mitts by using needles a size larger or smaller.

C6B

Place next 3 sts on cn and hold in back, k3 sts, k3 sts from cn. If sts have already been picked up onto cn, hold in back and complete the cable.

COWL

CO 128 sts. Do not join yet.
Knit 8 rows.
Next row: K4, then turn the left-hand needle counterclockwise one full turn around the first 8 rows (this will twist the knit piece), k4 sts and twist again. Continue to k4 and twist the knitting to the end of the row.

Now pm, join, and continue to work the remainder of the cowl in the round.

Knit 1 round.

Begin cable pattern as follows:
Rnd 1: *C6B, yo, k2tog; repeat from * for a total of 16 repeats.
Rnds 2 and 3: Knit.
Rnd 4: Knit to 3 sts before m; place the 3 sts on cn and hold in back.
Rnd 5: Sl m, finish working C6B, yo, k2tog, *C6B, yo, k2tog; repeat from *.
Rnds 6 and 7: Knit.
Rnd 8: Knit to 3 sts before m; place these 3 sts on cn and hold in back.

Repeat Rnds 5–8 eight more times or until desired size. End on Rnd 8. Last Rnd: Sl m, finish working C6B, k2; *C6B, k2. Repeat from * to m.

Remove m and proceed with a picot bind off as follows:

*CO 2, BO 5, slip loop from right needle back onto left needle. Repeat from * until all sts are bound off. Cut yarn and pull end through last loop.

Sew the first 9 rows together and weave in ends.

MITTS

Note: This pattern is written to be knit flat and seamed. It can also be easily adapted for knitting in the round. To do this, see double-pointed needle modifications at the end.

CO 36 sts.

Knit 8 rows.
Next row: K4, then turn the left-hand needle counterclockwise one full turn around the first 8 rows (this will twist the knit piece), k4 sts and twist again, continue to k4 and twist the knitting to the end of the row.

Next row (WS): (P11, M1) twice, p to end—38 sts.

Begin pattern for left mitt *(right mitt directions are in italic when they differ from the left)*:

Row 1: K across to last 8 sts. Pm on right needle and C6B, yo, k2tog. *(K 12 sts, pm on right needle and C6B, yo, k2tog, knit remaining sts.)*

Rows 2 and 4: Purl.

Row 3: Knit.

Row 5: K to 3 sts before m; place those 3 sts on cn and hold in back, sl m, finish working C6B, yo, k2to, k to end.

Rows 6 and 8: Purl.

Row 7: Knit.

Rows 9–16: Repeat rows 5–8 two times.

Row 17: Knit 14 sts, M1, k1, M1, k to 3 sts before m; place those 3 sts on cn and hold in back, remove m, finish working C6B, yo, k2tog, knit to last 8 sts, pm on right needle and C6B, yo, k2tog. *(C6B and remove m, yo, k2tog, k4, pm on right needle, C6B, yo, k2tog, k3, M1, k1, M1, knit to end.)*

Row 18: Purl.

Row 19: K14 *(23)* sts, M1, k3, M1, knit to end.

Row 20: Purl.

Row 21: K14, M1, k5, M1, k to 3 sts before m; place those 3 sts on cn and hold in back, sl m, finish C6B, yo, k2tog, knit to end. *(Knit across to 3 sts before m, place those 3 sts on cn and hold in back, sl m, finish working C6B, yo, k2tog, cont knitting until there are a total of 23 sts on right ndl, M1, k5, M1, knit to end.)*

Row 22: Purl.

Row 23: K14 *(23)*, M1, k7, M1, k across.

Row 24: Purl.

Row 25: K14, M1, k9, M1, knit across to 3 sts before m; place those 3 sts on cn and hold in back, sl m, finish working C6B, yo, k2tog, knit to end. *(Knit across to 3 sts before m; place those 3 sts on cn and hold in back, sl m, finish working C6B, yo, k2tog, continue knitting until there are a total of 23 sts on right ndl, M1, k9, M1, knit to end.)*

Row 26: Purl.

Row 27: K14 *(23)*, M1, k11, M1, knit across.

Row 28: Purl.

Row 29: K14, M1, k13, M1, knit across to 3 sts before m; place those 3 sts on cn and hold in back, sl m, finish working C6B, yo, k2tog, knit to end. *(Knit across to 3 sts before m; place those 3 sts on cn and hold in back, sl m, finish working C6B, yo, k2tog, cont knitting until there are a total of 23 sts on right needle, M1, k13, M1, knit to end.)*

Row 30: Purl.

Row 31: K13 *(22)*, put 17 sts on piece of scrap yarn, CO 3 sts to right needle, continue to knit across—38 sts on the needle.

Row 32: Purl.

Row 33: Knit across to 3 sts before m; place those 3 sts on cn and hold in back, remove m, finish working C6B, yo, k2tog, knit to end. *(Place first 3 sts on cn and hold in back, remove m, finish working C6B, yo, k2tog, knit to end.)*
Row 34: Purl.
Row 35: Knit.
Row 36: *(P11, p2tog)* two times, purl to end—36 sts.
Rows 37–44: *K2, p2, repeat from * for top ribbing.

BO all sts.

THUMB

Worked same for both hands; worked in the round.

Put 17 sts from waste yarn on dpn and pick up 3 sts along CO sts between thumb and hand, spreading sts evenly onto 3 dpns—20 sts.

Rnd 1: Knit.

Rnd 2: Knit around, dec 4 sts total by using k2tog, mostly in area between thumb and hand (you can place decreases as needed to cinch up holes that might begin to form in this area)—16 sts.

Rnds 3–9: *K2, p2, repeat from * for ribbing.

Bind off rem sts.

Sew mitt together at side seam.

DOUBLE-POINTED NEEDLE MODIFICATIONS

The sample mittens for this pattern began being knit in the round, but because of pooling of colors in this particular yarn, I chose to knit them flat instead. This way the colors looked more like they do in the cowl.

If you want to make the mitts in the round, begin this way:

CO 38 sts, do not join yet.

Knit 8 rows.

Next row: K4, then turn the left-hand needle counterclockwise one full turn around the first 8 rows (this will twist the knit piece). K4 and twist again. Continue to K4 and twist the knitting to the end of the row.

Now pm, join, and continue the remainder of the mitts in the round (so, if it says to work across, now you will be working around).

Knit 1 rnd.

Begin patt for left mitt *(right mitt directions in italic)*.

Work mitts the same as for flat—*except* all rnds are knit with RS facing. Be sure to leave m that denotes beg of the rnd in place and place additional ms as noted for the cable twists. Work thumbs the same as for flat mitt.

2007
2006
2005

IN THE SUGAR BUSH

IN LATE WINTER I achingly look for signs of spring. One February evening I hear the maniacal courtship song of barred owls. In early March, before the arrival of robins and crocuses, I spy my first bluebird, scouting out the nest box on the fence post outside my kitchen window. And then almost overnight nearly every roadside maple down in Shelburne Center is festooned with a pair of galvanized buckets, marking the start of sugaring season. Sugaring starts down in the valley and works its way up into the hill towns. Once that's under way, I know we've turned a corner. Spring shearing and lambs will soon follow.

No one watches late-winter weather as closely as the maple farmers. A string of warm days in the midthirties, with nights below freezing, starts the sap run in the sugar maples. Every farm has taps in a sugar bush and a shed for boiling sap. Farmers start tapping in late February or early March.

Setting roadside taps is easier than getting into the woods, unless the snow is granular and compacted. Many farms still collect sap in the traditional way, by drilling holes in the tree, tapping in a tap, and hanging galvanized buckets.

But many farms have modernized and now attach a network of plastic tubing to string together long runs of trees in a stand. Gravity feeds the sap from many trees to a main line that runs to a large holding tank that can easily be reached by truck or Jeep for collecting.

Sugar maples (*Acer saccharum*) are also known as hard maples or rock maples, since they like growing on steep rocky slopes. It's easy to confuse them with their cousins, the soft maple or the red maple, especially in winter, when the trees are naked of defining foliage. The bark of sugar maples is vertically alligatored. Its buds are tight and spear-like against the winter sky.

The Davenport family has been sugaring on the Patten for five generations. Norm's father, Russ Davenport, grew up farming as did his father before him. In the time of Russ's boyhood the process of sap collection from buckets hanging on maple trees happened using horses pulling a sled with a wood-stave tank chained to it. It was slow going with horses in deep snow. He remembers collecting sap this way in 1935 when he was six years old. His grandfather handed him the reins one day and said, "You're the boss." When the family went from

using horses to using a crawler tractor in 1950, he remembers thinking, "I could do everything with this!"

In Russ's lifetime the sugaring industry continued to make labor- and time-saving advances—from collecting sap from buckets tree by tree in the woods to using a plastic pipeline strung from tree to tree and feeding into a large collection tank. In the sugarhouse at their farm the tap lines run from the woods above their home on the mountain right into a collection tank at the side of the sugarhouse. One glance at the clear pipe near the ceiling of the sugarhouse tells Norm and his dad if the sap is running.

From between forty to forty-five thousand individual taps, the Davenports produce more than eight hundred gallons of syrup in a season. It takes forty gallons of maple sap to produce one quart of syrup through the process of boiling off the water in the reverse-osmosis evaporator at the sugarhouse. If there's steam pouring from the roof of the sugarhouse, you know the Davenports are boiling syrup.

At the edge of my hayfield, the slope would allow the main line to run right down the hill to a 250-gallon tank at the road. This stand hadn't been tapped

in more than a decade. A cluster of young sugar maples are offspring of three massive 250-year-old maples that tower above. Planted randomly by the prevailing west wind at the field's edge, this crowded young stand will eventually have to be thinned. Mature maples line the stone wall of the sheep's sugar bush pasture. In summer the sheep tuck themselves into the shade of these trees to get out of the sun. I occasionally find old bits of tubing in this pasture, remnants from the farm's previous owners. I pull the tubing from the ground wherever I find it so the sheep won't get tangled in it. Taps from these trees will connect to our main line at the lane and will share the roadside collection tank with the young stand. Once the taps and lines are set, gravity will do most of the work.

Fred Davenport comes up after morning chores to give me a lesson in tapping. Using a battery-powered drill, he bores holes about two inches into the trunk. Angling the bit upward helps the sap flow down and out. In the early season, tap holes are drilled on the warmer south-facing sides of trees. Since we are now midseason, we tap the cooler side, in hopes of extending the run. The number of taps is determined by the diameter and age of the tree. The general rule is not to tap anything less than ten inches in diameter. Fred sets taps above a root or below a healthy branch, since that's where the sap flows. If the tree has been tapped in previous years, you bore two inches to either side and a foot above or below the earlier tap holes.

A black plastic spout fits easily into the perforated bark. Energetically, I tap in each spout with a small hammer. Fred reminds me (more than once) that the goal isn't to drive the entire spout right into the tree. Just tap lightly until the hammer bounces off the end of the spout; then you're good. I eventually get the feel of it.

Working with Fred, I learn interesting things about the challenges of sugaring. Earlier that morning he had been tapping down in Howard's Hole, not far from our farm. For that stand, a gasoline-powered pump runs a vacuum line that brings the sap uphill to a collection tank on the back of a truck. Woodpeckers sometimes screw up the system. They mistake the faint whistling sound made by the vacuum for insects and peck holes in the tap fittings. I laugh, picturing an army of woodpeckers sabotaging the Davenports' taps.

A week earlier, Norm and I had measured how many feet of tubing we would need to bring the main line from the farthest corner of the sugar bush to the road, taking a rough count of the number of taps we would need along the way. Starting at the road, we walked uphill, unreeling nearly seven hundred feet of three-quarter-inch tubing from a giant wooden spool. This is the main line. Norm says it's a good idea to limit the number of splices into the main line, so we work to hook up as many trees as possible to a series of feeder lines. Where feeder lines tie into the main pipeline, he makes connections, punching a small hole into the main tube with a special tool and adding a fitting gasket. The feeder line slips over the fitting.

The random arrangement of our young trees calls for creative interlacement. Soon the feeder lines snake like tributaries from tree to tree and then flow into to the main line. It's an exercise in connecting the dots, stringing feeders to link as many trees as possible, always starting with the tree lying farthest uphill and working our way down toward the main line.

The "drops" are the small leaders fitted with a tapping spout at the end that attaches to the tree and a T fitting at the other end, where the leader ties into the feeder line. Tight fittings and minding the slope are important. The sap flows easily downhill, but if the line sags, it slows the entire system.

We walked the entire line, tightening the feeder lines wherever they bellied, adjusting tension.

While bracing myself against a stone wall and tapping a spout on the steepest part of the slope, I view Mount Monadnock's snowy peak framed by blue sky and the silvery trunks of maple trees. Although I see this mountain every day from the house, I've never seen it from this vantage point. My sheep barn, a quarter mile up the road, looks small from this perspective.

The last step was to put the 250-gallon collection tank in place near the road at the end of the run. From here the sugar bush looks like a cat had its way with a giant ball of yarn among the trees. With any luck, the temperatures will drop below freezing at night and jump-start the sap run.

Back at the Davenports' sugarhouse, the evaporator is fired up. Norm will be boiling sap each day from now until the buds swell on the trees. Then the season ends and it's time to disconnect the tubing and wash and store it for the next season. But until then, we'll be at the Davenports' Sugar House Restaurant every Sunday morning for pancakes.

SUMAC MITTENS

Designed by Kate Gilbert

Kate's tightly knit mittens add warmth, charm, and a dash of color to a cold winter's day. The pattern—a leaf mosaic on the back of the mitten and a diamond motif on the palm—celebrates color in a season of mostly gray and white.

Knit from a decadent blend of Cormo wool, alpaca fiber, and Bombyx silk, the mittens have an itch factor of zero—and are much too pretty to be worn for barn chores.

Finished Size

8" (20 cm) hand circumference and 9¼" (23.5 cm) long (above the I-cord edge)

Yarn

- Foxfire Fiber & Designs Cormo Silk Alpaca (70% Cormo, 20% alpaca, 10% silk; 190 yd [173 m]):1 skein main color (MC); 1 skein contrast color (CC). Shown in Aster (MC) and Winterberry (CC).

Needles

- One set of double-pointed needles size US 3 (3.25 mm)
- One set of double-pointed needles size US 5 (3.75 mm)

Notions

Scrap yarn to be used as stitch holder, stitch marker

Gauge

32 sts and 36 rows = 4" (10 cm) in colorwork pattern on larger needles. Adjust needle size as necessary to obtain gauge.

KEY
K1 MC
K1 CC
K2tog with color indicated
SSK with color indicated
K3tog with color indicated
SSSK with color indicated
THUMB CHART
20
18
16
14
12
10
8
6
4
2
MITTEN CHART
82
80
78
76
74
72
70
68
66
64
62
60
58
56
54
52
50
48
46
44
42
40
38
36
34
32
30
28
26
24
22
20
18
16
14
12
10
8
6
4
2

FOUR-STITCH I-CORD EDGING

Using larger dpns, CO 4 sts using scrap yarn. Do not turn. Slide sts to other end of needle, k4 in scrap yarn. Join MC, *k4, *do not turn,* slide sts to other end of needle; rep from * for 63 rows total. Cut yarn, leaving a tail of at least 12" (30 cm). Take out scrap yarn and put sts on a dpn. Graft I-cord together into a circle, avoiding twisting it. Weave in ends.

MITTEN

With MC and smaller dpns, pick up 64 sts evenly around the I-cord.

Place marker, join CC and start working in rounds. Work mitten according to chart, until last st of rnd 38 for left mitten (last stitch of rnd 37 for right mitten).

THUMB SETUP

When thumb opening sts are reached, k9 sts in scrap yarn, sl 9 sts back to left needle, k9 with MC, carrying CC across.

Work rem rnds of mitten to end of chart working decs as charted.

Graft sts at tip together. Weave in ends.

THUMB

Remove scrap yarn and place lower 9 sts on dpn 1 and the upper 9 sts on dpn 2 (using smaller dpns).

Rejoin yarns, and working according to rnd 1 of the Thumb Chart, pick up and knit 3 sts at side of Thumb, work across dpn 1, pick up and knit 3 sts at side of Thumb, work across dpn 2—24 sts. Place marker and redistribute sts onto 3 dpns.

Work rem rnds of thumb to end of chart, working decs as charted—3 sts rem.

Pass yarn through last 3 sts and pull tight.

Weave in ends. Secure CC yarn that was carried across thumb opening.

BINDING OFF

THE LAST ROW in a knitting project is the final interlacement that secures the piece and creates a finished edge. When the last stitch has been passed and the tail woven in, knitters have a moment of what I call "complete completion." Wash and block and send this sweater out into the world. I feel the same way whenever I wind newly spun yarn off my bobbin and onto a skein winder, or when I cut a finished piece of woven cloth from my loom. There's a moment of excitement and accomplishment.

Unlike most knitting projects, fiber farming is a perpetual WIP (work in progress). Although there are definite events that feel like casting on and starting anew (shearing day, lambing season, planting the garden), there is no one moment of complete completion. In that respect, the farm is a perpetual knit in the round. The beginning of a row is determined by wherever I place my stitch marker. Some years the gauge is perfect and the pattern is spot-on. In others we twist and drop stitches. In some years the glitches are barely noticeable. At other times, they leave big holes. And so I rip back to good and start again.

As a knitter I am notorious for not finishing projects. But I do get a sense of complete completion whenever a knitter shares a finished project created from my yarn or fiber. It's a moment of resolution as I reflect on all that has gone into completing the circle: grass, fields, sheep, goats, lambs, rams, llamas, tractors, fence, shearing, mills, wool, roving, yarn, dye pot, knitters, needles.

I've no doubt our farm will continue to evolve over time. For now, I'm binding off. It's time to head down to the barn to feed the sheep.

ACKNOWLEDGMENTS

This book came to fruition with the help of many people to whom I am deeply grateful.

Many thanks go to my agent, Linda Roghaar, and to my editor, Jennifer Urban-Brown of Shambhala Publications, for believing in this project and for their patient encouragement throughout the entire process.

Thank you, Ben Barnhart, for enduring many odd hours involved in capturing my farm in beautiful photography.

Thank you to the designers, Barbara Giguere, Kate Gilbert, Christa Giles, Lisa Lloyd, Marnie MacLean, Melissa Morgan-Oakes, and Holly Sonntag, for conceiving lovely designs for my yarn, and to chef Margaret Fitzpatrick for creating tasty recipes using the food we raise on our farm.

I deeply appreciate the friends and acquaintances who generously shared their time and expertise: Bruce Clement, Fred Davenport, Lisa and Norman Davenport, Russ Davenport, Temple Grandin, Ivy Palmer, John and Mary Ellen Warchol, and Christopher Zinn.

Many thanks to my test knitters: Teresa Campbell, Barbara Giguere, Valerie Heuchan, and Trisha Maryea; and to Kate Corriero, Julia Denig, and Mia Tabery Monseratte for modeling knitwear projects.

Thank you to my customers, especially the members of our farm's yarn and fiber CSA, Sheep Shares, for your friendship and loyalty.

I am indebted to my mentors in sheep husbandry, agriculture, and fiber craft: Lisa and Norm Davenport, Alice Field, Chris Hamel, Bob Ramirez, Andy Rice, and Jean and Merle Willmann.

This project would not have been possible without the support of friends and family throughout the process: Dorothy Aquavella, Barbara Bell, Amy Gordon, Caleb Kissling, Jim Noonan, Clara Parkes, Caleb and Kundan Parry, Holly Sonntag, Kathryn Greenwood Swanson, Kim Keefe Swasey, and Midori Tabery.

My love and thanks go to my sisters—Trish and Kathleen—for helping launch Foxfire Fiber & Designs and for logging so many hours in my booth at sheep and wool festivals; and to Mom and Dad, for encouraging the farmer in me by letting me raise rabbits in the backyard as a child.

I send my deepest love and gratitude to my husband, Mike, for understanding how much this project means to me, for shouldering the load when I've been in way over my head, and for giving me the space and time to write.

ABBREVIATIONS

approx approximately
beg beginning
BO bind/bound off
CC contrasting color
cn cable needle
CO cast on
cont continue
dec decrease
dpn(s) double-pointed needle(s)
inc increase
K or k knit
k2tog knit 2 stitches together
MC main color
m(s) marker(s)
M1 Make one additional stitch by lifting the bar between the last stitch and the next stitch onto the left needle and knitting into the back of this stitch.
M1L Make one left: insert left needle from front to back into the horizontal bar between the last stitch worked and the next stitch and knit (or purl) the bar through the back.
M1R Make one right: Insert left needle from back to front into the horizontal bar between the last stitch worked and the next stitch and knit (or purl) the bar through the front.
ndl(s) needle(s)
no. number
P or p purl
patt(s) pattern(s)
pm place marker
p2tog purl 2 stitches together
rem remain/remains/remaining
rep(s) repeat(s)
rnd(s) round(s)
RS right side
sl slip
ssk slip first stitch as to knit, slip next stitch as to knit, then knit these 2 stitches together
st(s) stitch(es)
St st stockinette stitch: knit all sts on RS rows and purl all sts on WS rows
tbl through back loop
w&t wrap and turn: Bring yarn to front of work between needles, slip next st to right-hand needle, bring yarn around this st to back of work, slip st back to left-hand needle, and turn work to begin working back in the other direction.
wpi wraps per inch
WS wrong side
yb yarn back
yf yarn forward
yo yarn over

GLOSSARY

blood grade: a system once commonly used to describe wool market grades in six categories: fine ½ blood, ⅜ blood, ¼ blood, low ¼ blood, common, and braid. The fractions refer to the percentage of Merino in the breed's genetic makeup (because so many breeds evolved from a crossing with Merino).

blow: a pass of the shearing blades. An efficient shearer can shear with a minimum number of blows (fifty to sixty blows per average-sized sheep is typical).

Bradford count: a method of measuring grade based on the number of 560 yard hanks that can be spun from one pound of clean wool.

braid: a term from the blood grade system used to describe a very coarse fleece.

broomie: the person responsible for sweeping the shearing board in between sheep.

crimp: the corrugated or wavy pattern of growth of a fleece.

crutching: shearing wool from a sheep's hindquarters often done in advance of breeding or lambing season or for hygienic purposes.

cuticle: the outer layer of a wool fiber, comprised of overlapping scales.

dags: manure-tipped wool.

drop spindle: a shaft with a whorl attached; the most basic tool used for hand spinning yarn.

drum carder: a large cylinder or drum lined with steel teeth that teases apart and opens wool locks in preparation for spinning.

fleece: a year's wool growth shorn from a sheep.

fleece-o: the person who picks up the freshly shorn fleece from the shearing board either to bag it or to toss it on the skirting table.

fribs: areas of a fleece (such as the topknot, leg, and belly wool) that are generally short and unusable.

grade: the relative fineness of wool fibers.

grist: the thickness of a strand of yarn, determined by the diameter of the drafted roving prior to receiving twist.

hand cards: paddles with wooden handles attached and whose faces are covered with cloth

lined in rows of short steel teeth. These fiber preparation tools are used in pairs to open up locks of wool for spinning.
hogget: a yearling sheep.
hogget fleece: the fleece of a hogget sheep. Also called "lamb's wool" if this is the sheep's first shearing and the fleece still has its milk tips (the original tips from birth).
keratin: protein that makes up wool fibers.
lanolin: a waxy substance naturally produced by the sebaceous gland in a wool follicle to provide a degree of water repellency for sheep. Sometimes called "grease," lanolin makes raw fleeces feel sticky, especially in fine wool sheep.
micron: a measurement of fiber diameter, 1/1000th of a millimeter, determined using a digital laser scan. This is the most common commercial measurement used in today's wool market.
rolag: a cylinder of carded wool created using hand cards in preparation for hand spinning.
roving: a fiber preparation for woolen or semiworsted-style yarn spinning. The fibers are carded into a random but open arrangement and then attenuated into one long continuous strand.
second cut: a second pass (cut) made in the same spot on a fleece, creating undesirable, very short pieces of wool.
semiworsted (style): method of spinning yarn when worsted-style spinning technique is used on carded roving (rather than on combed top).
sheepo: the person responsible for the flow of sheep in the shearing barn, especially filling the catch pen for the shearer.
skirting: the process of removing areas of wool of inferior quality from a fleece.
spindle: a tool used for spinning yarn. It can refer to a spinning wheel or drop spindle.
staple: the natural form in which a fleece separates. A staple may be blocky, tapered, or pencil-shaped.
teg: a two-year-old sheep (age 12–28 months).
thrums: the remnants of warp yarn on the loom after finished woven cloth has been cut from the loom; sometimes called "warp waste."
top: a fiber preparation for worsted-style yarn manufacture. The fleece is carded and then combed to align the fibers and to remove short pieces of fibers.
tup: a ram lamb old enough to be successfully used in breeding.
wether: a male sheep that has been castrated.
"wool away": a shearer's command to remove a shorn fleece from the board.
woolen (style): method of spinning yarn from carded roving (often fine wools of shorter length) to create a yarn that is lofty, bulky, and has a nap.
worsted (style): method of spinning yarn from combed top (fibers that are combed into alignment and longer in length) to create a yarn that is sleek and smooth.

GUIDELINES FOR SAFELY DYEING YARN AND FIBER

The projects described in this book are for use on protein fibers using weak acid dyes set with vinegar (or citric acid) and heat. Dyeing yarn and fiber is an exciting, creative process, but safety guidelines must be heeded.

Always consult the dye supplier's instructions first for safety precautions in using their textile dyes and auxiliary materials.

- Wear rubber gloves to protect your skin when handling dye products.
- When mixing dye powders into solution, protect your lungs (and those of any others working nearby) by wearing a particle filter mask.
- Make sure your work area has good ventilation. Whenever possible, do your dyeing outdoors.
- If working indoors, wear an MSHA/NIOSH respirator mask with cartridges for dust, mists, and fumes to protect your lungs from dye powder and dye-bath vapors.
- If you experience an adverse reaction to any dye or dye auxiliary material, move away from the area and into fresh air. If symptoms persist, disuse the product and seek medical attention.
- Wear safety glasses to protect your eyes.
- Do not eat, drink, or smoke while dyeing.
- Keep dye pots and utensils separate from food utensils. Once you've used something for dyeing, do not use it for food preparation.
- When mixing dye powder into liquid, cover your work area with dampened newspapers or paper towels.
- Do not mix powders near a furnace or air intake pipes.
- Cover work surfaces with a plastic table covering (shower liners work well for this) and wipe up all spills immediately.
- Keep children and pets out of the room or dye area for their safety.
- If pregnant or nursing, consult manufacturers' or distributors' instructions before using dye products.
- Store dyes and auxiliary materials in a cool, dry place out of reach of children and pets.

- If storing unused dye solution, be sure to label containers clearly and store all dyes out of reach of children and pets.
- Always add acid (vinegar) to water, not the other way around.
- Always exhaust your dye bath. Use an appropriate amount of dye for the amount of fiber so that all dye in the pot is absorbed. Balance the pH of dye pot water by using baking soda if necessary before disposal.
- For complete instructions and more projects for dyeing yarns and fibers, please consult my book *Teaching Yourself Visually Hand-Dyeing* (Wiley Publishing, 2009).

RESOURCES

STUDIO SUPPLIES AND MATERIALS

Dharma Trading Company (source for Dharma Acid Dyes and Jacquard Acid Dyes) www.dharmatrading.com

Louet North America (source of Gaywool Dyes from Australia) www.louet.com

PRO Chemical & Dye (source for PRO WashFast Acid Dyes and PRO One Shot Dyes) www.prochemical.com

Stockbridge Herb Farm, South Deerfield, Mass.; John and Mary Ellen Warchol, South Deerfield, Mass.: culinary herbs

W. Cushing & Company (source of Cushing Perfection Dyes) www.wcushing.com

BOOKS

Barnard, Ellsworth. *In a Wild Place: A Natural History of High Ledges*. Massachusetts Audubon Society, 1998.

Bowen, Godfrey. *Wool Away: The Art and Technique of Shearing*. Van Nostrand Reinhold Company, 1955.

Davenport, Russell M. *The Sunny Side of Mt. Massaemet: The Recollections of Russ Davenport*. Self-published, 2011.

Drummond, Sue Black. *Angora Goats the Northern Way*. Stony Lonesome Farm, 1988.

Grandin, Temple. *Humane Livestock Handling: Understanding Livestock Behavior and Building Facilities for Healthier Animals*. Storey Publishing, 2008.

Parry, Barbara. *Teach Yourself Visually: Hand-Dyeing*. Wiley Publishing, 2009.

Rezendes, Paul. *Tracking and the Art of Seeing: How to Read Animal Tracks and Sign*. Harper Perennial, 1999.

Simmons, Paula. *Raising Sheep the Modern Way*. Storey Publishing, 1998.

Stove, Margaret. *Merino: Handspinning, Dyeing and Working with Merino and Superfine Wools*. Interweave Press, 1991.

Teal, Peter. *Hand Woolcombing and Spinning: A Guide to Worsteds from the Spinning-Wheel*. Blandford Press, 1976.

ABOUT THE DESIGNERS AND CONTRIBUTORS

Ben Barnhart—Ben is a freelance commercial and editorial photographer living in Conway, Massachusetts. His photographs have appeared in the *New York Times*, the *Boston Globe*, *Yankee* magazine, *Boston* magazine, and many other publications, books, websites, and annual reports. Among his commercial clients are colleges and universities in the Pioneer Valley including UMass Amherst, Mount Holyoke College, Springfield College, and others. He also teaches in the online journalism program at UMass and holds a bachelor's degree in journalism from Boston University. Visit www.bbimages.com to see his work and learn more.

Margaret Fitzpatrick—A native of New York City, Margaret has been a working chef and restaurant owner in the Pioneer Valley for more than twenty-two years. Chef Fitzpatrick owned one of Western Massachusetts's premier catering companies as well as hosted cooking shows on public television and local cable networks. A true believer in the farm-to-table movement, she is a graduate of the Culinary Institute of America and has studied abroad in France.

Barbara Giguere—Barbara is a married mother of seven and grandmother of three who raises sheep and alpacas on a farm in Shelburne, Massachusetts, less than three miles from Springdelle Farm. An accomplished spinner, knitter, and sewer, her designs have won many awards at local fairs. She markets her own yarns under her label Dragon Brook Yarns.

Kate Gilbert—Kate is the editor-in-chief of *Twist Collective*, an online magazine about knitting. She's mother to the coolest kid in Montreal and runs, bakes, and reads in her spare time. Find her newest designs at twistcollective.com.

Christa Giles—Christa has contributed knitwear designs to *Twist Collective*, *Knitscene*, and *Fresh Designs*, and self-publishes at christaknits.com. She teaches weekly classes in knitting and modern hoop dance, and enjoys cycling around Vancouver, British Columbia, on her custom bike, Ruby.

Lisa Lloyd—Lisa's book, *A Fine Fleece*, was one of the first knit-design books to feature handspun, farm-sourced yarn in each and every design. Her work has appeared in *Interweave Knits*, and she has been a contributing editor to *Wild Fibers* magazine.

Marnie MacLean—Marnie has been knitting since she was six years old and designing since 2003. She lives in beautiful Portland, Oregon, with her husband and three fantastic dogs. Marnie is a regular designer and production assistant for *Twist Collective* and also has published independently on her own site. You can see many more of her free and for-sale patterns at http://marniemaclean.com.

Melissa Morgan-Oakes—Melissa is a knitting instructor, pattern designer, and author of *2-at-a-Time Socks*, *Toe-Up 2-at-a-Time Socks*, and *Teach Yourself Visually Circular Knitting*. She sells self-published patterns on Ravelry and Craftsy, and travels around the country visiting yarn shops and teaching knitting techniques to eager audiences.

Holly Sonntag—Holly's knitwear design for this book evolved out of necessity. Many of my blog readers have offered to knit for our lambs—and thanks to Holly, now they can. Holly became a fiber lover through working with the animals that grow fiber. Her sweater brings fiber full circle, providing warmth for newborn lambs in the same way their wool will one day provide warmth for us.

ABOUT THE AUTHOR

Barbara Parry's life revolves around her family, her golden retrievers Farley and Zoe, and the sheep, llamas, donkeys, and goat who live at Springdelle Farm. When not tending sheep Barbara writes, weaves, knits, and gardens. She loves sharing what she has learned from sixteen years of working with sheep and fiber—serving as a consultant to start-up fiber farms and teaching fiber arts skills. For more stories about her farm, visit her blog *SheepGal: Adventures in Yarn Farming* at www.SheepGal.com.

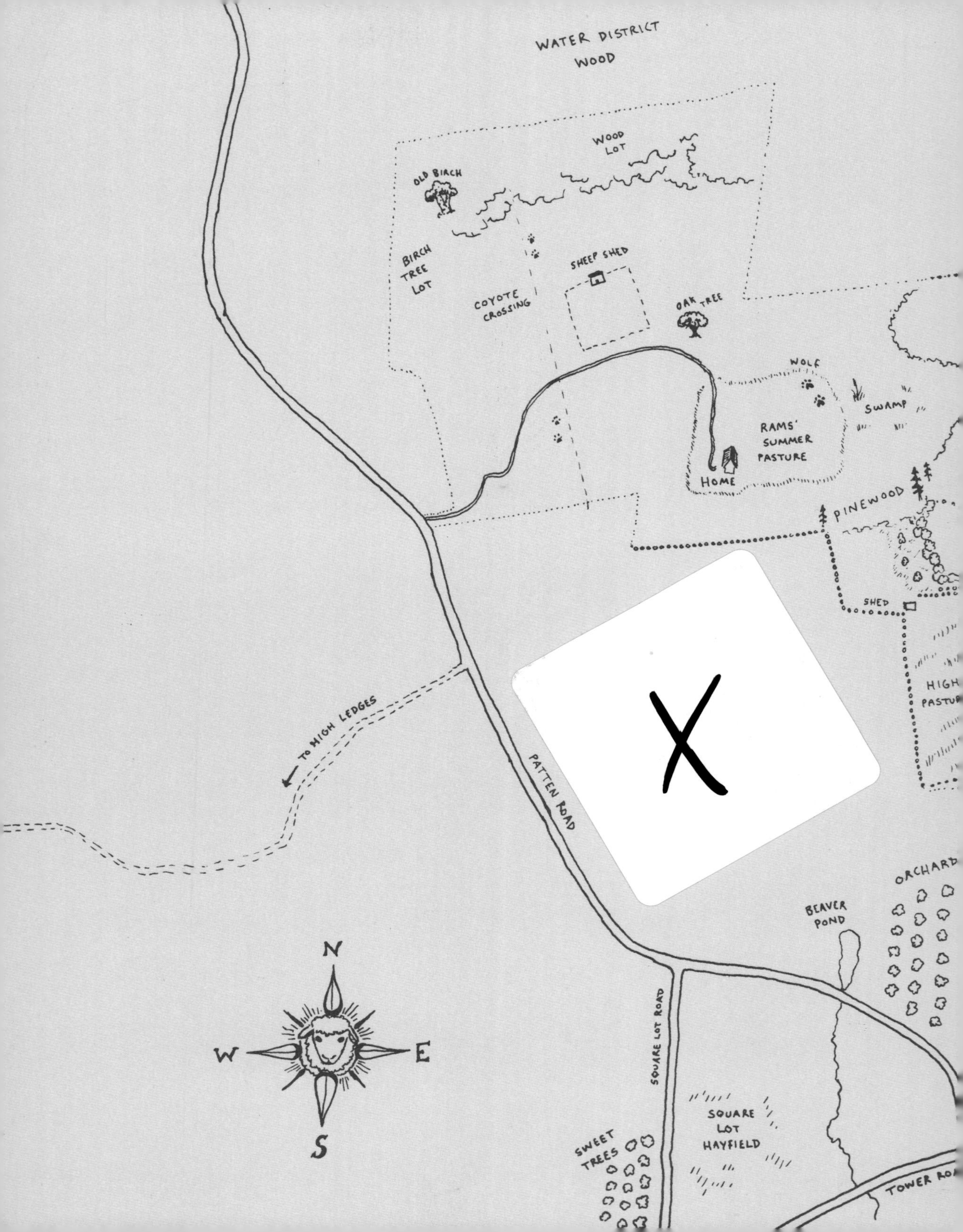
WATER DISTRICT WOOD
WOOD LOT
OLD BIRCH
BIRCH TREE LOT
COYOTE CROSSING
SHEEP SHED
OAK TREE
WOLF
SWAMP
RAMS' SUMMER PASTURE
HOME
PINEWOOD
SHED
HIGH PASTU
TO HIGH LEDGES
PATTEN ROAD
X
ORCHARD
BEAVER POND
N
W
E
S
SQUARE LOT ROAD
SQUARE LOT HAYFIELD
SWEET TREES
TOWER ROA